1976

1
APPLIED DIFFERENTIAL EQUATIONS

VNR NEW MATHEMATICS LIBRARY

under the general editorship of

J. V. ARMITAGE
Professor of Mathematics
University of Nottingham

N. CURLE
Professor of Applied Mathematics
University of St. Andrews

The aim of this series is to provide a reliable modern coverage of those mainstream topics that form the core of mathematical instruction in universities and comparable institutions. Each book deals concisely with a well-defined key area in pure or applied mathematics or statistics. Many of the volumes are intended not solely for students of mathematics, but also for engineering and science students whose training demands a firm grounding in mathematical methods.

APPLIED DIFFERENTIAL EQUATIONS

N. CURLE
Professor of Applied Mathematics
University of St. Andrews

VAN NOSTRAND REINHOLD COMPANY
LONDON
NEW YORK CINCINNATI TORONTO
MELBOURNE

VAN NOSTRAND REINHOLD COMPANY
Windsor House, 46 Victoria Street, London, S.W.1

INTERNATIONAL OFFICES
New York   Cincinnati   Toronto   Melbourne

Library of Congress Catalog Card No. 70–171372

First Published 1972

Printed in Great Britain by
Butler & Tanner Ltd, Frome and London

Contents

Preface

This book is based upon material that has been tried and successfully proven in many years of university teaching. This approach to differential equations lays emphasis not only upon how certain types of ordinary differential equations may be solved, but also on the manner in which these equations might possibly arise in practice, and the practical significance of the solutions, once obtained. This is the way things happen in real life, and I believe it to be the way in which the subject should be taught.

Broadly speaking, the level of the book should be appropriate to first-year students at British universities in mathematics, astronomy, chemistry, physics, and in the various branches of engineering. The precise amount of ground which might be covered in a given course will naturally depend on the circumstances. At St. Andrews we have a first-year course of approximately 17 lectures (plus tutorials) in which we cover the whole book excluding Chapters 3 and 7. In universities where mathematicians are taught separately from non-mathematicians and in which Honours Degree and Ordinary Degree students are likewise separated at an early stage, a class of Honours mathematicians might well cover the whole book in about this period of time. For those who must needs proceed more gradually, Chapters 3 and 7 could probably better be dealt with at the start of a later course, when the student has attained a greater maturity.

It should perhaps be stressed that it is not intended that the student should be expected to understand the details of the ideas underlying the various problems studied, for example, in Chapter 4. In particular, no knowledge of rigid-body dynamics, and virtually no knowledge of particle dynamics is required. On the other hand the chapter should serve as a valuable first introduction to the notion of mathematical modelling. The idea of making simple approximations which contain the bare essentials of a problem, and which lead to fairly straight-forward mathematics, is one to which an early introduction is desirable. The refinements which are permitted by developing mathematical maturity will follow more readily if sound foundations are laid at an early stage.

I have aimed at including a realistic range of questions at the end of each chapter, and the ordering of these has been carefully arranged. The student who is not able to make a reasonable attempt at most of these questions needs to go back and study the text much more thoroughly. Some of these examples have been culled from examination papers at a number of universities, notably Manchester, St. Andrews, and Southampton, and I am pleased to acknowledge the permission given to use these. A number are my own.

St. Andrews
1971   N. CURLE

CHAPTER 1

Introductory Ideas

1.1 Definitions and notation

It is useful to begin a book on differential equations by asking precisely what a differential equation is, and in what way it differs from an algebraic equation. An algebraic equation is, of course, simply a relationship between two (or more) otherwise independent quantities. For example, the relationship

$$x^2 + y^2 = 1 \qquad (1.1)$$

is an algebraic equation in the two variables x and y. Suppose we differentiate this equation with respect to x. We then find, after division by two, that

$$x + yy' = 0.\dagger \qquad (1.2)$$

Equation (1.2) differs from equation (1.1) in that although, basically, it is still a relationship between x and y, this relationship now involves the differential coefficient, y', of y with respect to x. It is for this reason that the relationship is referred to as a *differential equation*.

When only two variables are involved, so that derivatives with respect to only one variable can arise, the equation is called an *ordinary* differential equation. On the other hand, if

† The standard notations for derivatives will be followed. Thus the derivatives of y with respect to x will be written as
either

$$\frac{dy}{dx}, \frac{d^2y}{dx^2}, \frac{d^3y}{dx^3} \cdots \frac{d^ny}{dx^n} \cdots,$$

or

$$y', y'', y''' \cdots y^{(n)}$$

1

there are more than two variables, and partial derivatives with respect to more than one variable arise, the equation is called a *partial* differential equation.

Definition. An ordinary differential equation (o.d.e.) is a relationship between variables x, y, and at least one of the derivatives y', y'', $y''' \ldots y^{(n)}$, of y with respect to x. Thus we may write

$$F(x, y, y', y'' \ldots y^{(n)}) = 0. \qquad (1.3)$$

The *order* of a differential equation is that of the highest derivative which appears therein. Thus

$$y' = 1 + y \quad \text{is of order one,}$$
$$y'' + 4y = x \quad \text{is of order two,}$$
$$(y')^2 - y' + 2y = x^2 \quad \text{is of order one,}$$

and

$$y''' + y^2 = 2x \quad \text{is of order three.}$$

The *degree* of a differential equation is the power to which the highest derivative appears in the functional relationship (1.3) above. The degrees of the above equations are accordingly one, one, two, and one, respectively. A first-degree equation is one which can be expressed so that the *highest* derivative appears linearly, that is, in the form

$$y^{(n)} = f(x, y, y', \ldots, y^{(n-1)}).$$

A *linear* equation, more restrictively, is one in which y and *all* of its derivatives (but not x) appear linearly. Thus a linear equation can be written in the form

$$a_n(x)y^{(n)} + a_{n-1}(x)y^{(n-1)} + \cdots + a_0(x)y = Q(x) \qquad (1.4)$$

It may be noted that, although a linear equation is automatically first degree, a first-degree equation is not necessarily linear.

1.2 Uniqueness of the solution of a differential equation

It is not difficult to realize that a differential equation by itself, with no subsidiary conditions, does not have a unique solution. For example, the extremely simple equation $y' = 0$ clearly has the solution $y = $ constant, where the constant can take any value whatsoever. Likewise the less trivial equation $y' = 1 + y$ may be shown to have the solution $y = ce^x - 1$, where c is a constant which takes any value whatsoever. Although, in a sense, it may seem odd that a precise equation does not appear to have a precise solution, we may surmise (by reference to the equation $y' = 0$) that the 'arbitrary constants' in the solution are really no more surprising than the 'constant of integration' in an indefinite integral.

Consider now the equally trivial, but more general, equation, $y^{(n)} = 1$. Let us find the solution by successive integration n times. Thus one integration yields

$$y^{(n-1)} = x + c_1,$$

where c_1 is the constant of integration. A second integration then yields

$$y^{(n-2)} = \tfrac{1}{2}x^2 + c_1 x + c_2.$$

A third integration yields

$$y^{(n-3)} = \tfrac{1}{6}x^3 + \tfrac{1}{2}c_1 x^2 + c_2 x + c_3$$

and, finally, subsequent integrations yield

$$y = \frac{1}{n!}x^n + \frac{1}{(n-1)!}c_1 x^{n-1} + \cdots + c_{n-1}x + c_n.$$

We note that the solution contains precisely n arbitrary constants, arising in the course of the n successive integrations. We are naturally led to consider the possibility, in fact true, which is contained within the following theorem. The theorem will be stated without proof.

Theorem. The solution of an nth order ordinary differential equation involves at most n arbitrary constants. Any solution of an nth order ordinary differential equation which contains n independent arbitrary constants is called the *general solution* of the equation. Any specific solution (obtained, for example, by assigning definite values to each of the arbitrary constants in the general solution) is referred to as a particular solution or a particular integral.

It should be noted that certain differential equations have solutions which cannot be derived from the general solution for any choice of values of the arbitrary constants. Such solutions are referred to as singular solutions. An example of a singular solution to a differential equation is found in the equation

$$x(y')^2 - yy' + 1 = 0. \qquad (1.5)$$

It may easily be verified, by direct substitution, that

$$y = ax + 1/a \qquad (1.6)$$

is a solution. Since this solution contains one arbitrary constant, and the equation is first order, it is accordingly the *general* solution. On the other hand it may equally be verified that

$$y^2 = 4x \qquad (1.7)$$

is a solution of the equation.

The existence of a singular solution, for this problem at any rate, can readily be given a geometrical explanation. The general solution (1.6) represents, for different values of the constant a, a set of straight lines. These lines happen to have an envelope, which is given by the solution (1.7). Apart from this observation, that singular solutions appear to arise when the curves represented by the general solution have an envelope, we shall not discuss them further here.

1.3 Boundary and initial conditions

In practice a differential equation will arise in conjunction with some sort of subsidiary condition or conditions. Take the following very simple example.

'An aircraft travels due north for 1 hour at an average speed of 500 kilometres per hour. What is its position at the end of the hour?'

It is rather obvious that the answer to the question depends upon precisely where the aircraft started from, but that it is unique once the *initial condition* is given.

It is clear that for an nth order equation, the n arbitrary constants of the general solution can be uniquely determined when n boundary or initial conditions are known—the term 'boundary condition' is used when the independent variable is a space co-ordinate and the term 'initial condition' when the independent variable is time.

1.4 The superposition rule for linear equations

From a practical point of view, linear equations are usually easier to solve than non-linear equations. The principal reason for this rests upon the fact that the general solution of a linear equation can be 'built up' in the manner which we shall now indicate. Consider the linear equation

$$a_n(x)y^{(n)} + a_{n-1}(x)y^{(n-1)} + \cdots + a_0(x)y = Q(x), \quad (1.8)$$

and the associated homogeneous equation

$$a_n(x)y^{(n)} + a_{n-1}(x)y^{(n-1)} + \cdots + a_0(x)y = 0. \quad (1.9)$$

Suppose that $y_1(x)$, $y_2(x)$, $\cdots y_n(x)$ are linearly independent† solutions of the homogeneous equation, and that $Y(x)$ is any solution of the inhomogeneous equation, i.e., a particular integral. Then the superposition rule is as follows.

Theorem

1. The general solution of the homogeneous equation (1.9), which solution is referred to as the complementary function (CF), is given by

$$y_{\text{CF}} = A_1 y_1 + A_2 y_2 + A_3 y_3 + \cdots + A_n y_n,$$

where A_1, A_2, ..., A_n are arbitrary constants.

† A number of functions $f_1(x)$, $f_2(x)$, $f_3(x)$ ... are said to be linearly dependent if it is possible to find numbers a_1, a_2, a_3 ... such that

$$a_1 f_1(x) + a_2 f_2(x) + \cdots = 0$$

for *all* values of x. For example, the functions x, x^2, and $x + x^2$ are linearly dependent, but the functions x, x^2, and $x + x^2 + x^3$ are linearly independent.

2. The general solution of the inhomogeneous equation (1.8) is the sum of the complementary function and a particular integral (PI).

Proof

1. Since y_1, y_2, y_3, etc., satisfy equation (1.9), we may write

$$a_n y_1^{(n)} + a_{n-1} y_1^{(n-1)} + \cdots + a_0 y_1 = 0,$$
$$a_n y_2^{(n)} + a_{n-1} y_2^{(n-1)} + \cdots + a_0 y_2 = 0,$$
$$\vdots$$
$$a_n y_n^{(n)} + a_{n-1} y_n^{(n-1)} + \cdots + a_0 y_n = 0.$$

We multiply these equations by A_1, A_2, ..., A_n, respectively, and add. It then follows that

$$a_n y_{\text{CF}}^{(n)} + a_{n-1} y_{\text{CF}}^{(n-1)} + \cdots + a_0 y_{\text{CF}} = 0, \quad (1.10)$$

so y_{CF} certainly satisfies equation (1.9). It must in fact be the general solution, since it contains precisely n arbitrary constants which multiply the n linearly independent functions $y_1(x)$, $y_2(x)$, ... $y_n(x)$.

2. If we note that the particular integral satisfies equation (1.8), it follows that

$$a_n Y^{(n)} + a_{n-1} Y^{(n-1)} + \cdots + a_0 Y = Q(x). \quad (1.11)$$

Upon adding equations (1.10) and (1.11) it follows that $y_{\text{CF}} + Y$ is also a solution of equation (1.8). Since n arbitrary constants are involved this is accordingly the general solution.

The practical utility of these results will be illustrated in due course. The essential point to note is that only one solution of the full equation (1.8) is required, and n solutions of the homogeneous equation (1.9), which may well be easier to obtain.

Tutorial Examples

1. State the order of each of the following ordinary differential equations. Which of them is linear? If any of the equations is not linear, state its degree.

(a) $y' = \cos x$.

(b) $y''' + 4y = 0$.

(c) $x^2 y'' + 2e^x y' = (x + 2)y$.

(d) $x^2 yy'' + \log x = 2xy^2$.

2. Differentiate each of the following equations as many times as may prove necessary, and then eliminate the constants a, b, c, from the resulting set of equations. Thus deduce the differential equations of which these are the general solutions.

(a) $y = a e^{-x} + b \sin x$.

(b) $y^2 = ax - x \log x$.

(c) $y = a e^x + b e^{2x} + c e^{-x}$.

State the order and degree of each of the differential equations.

CHAPTER 2

Equations of the First Order and First Degree

We have already seen that a first-order equation is one in which y' is the highest derivative occurring. Such an equation is also first degree if y' appears linearly. Accordingly, a first-order first-degree equation is one which can be written as

$$y' = F(x, y). \qquad (2.1)$$

There will normally be no difficulty in integrating this equation numerically. Thus, with a given starting point, $y = y_0$ when $x = x_0$, we use equation (2.1) to calculate the value of $y'(x_0)$. Then the value of y corresponding to a nearby value of x, say $x = x_0 + h$, will be approximately equal to $y_0 + hy'(x_0)$. This gives us a new starting point for the next step of the integration procedure, and so on.

The detailed refinements of this process will not be discussed here, but it is well worth noting that a numerical solution can be found for any reasonable form of the function $F(x, y)$. If the boundary conditions are known, and if the equation itself does not contain a parameter such that solutions are potentially required for a range of values of the parameter, then only one solution is sought, and this will often best be obtained by numerical methods. On the other hand, if the solution is required for each of several values of a parameter, it is of considerable advantage when an explicit solution can be found in terms of either known or readily tabulated functions. We shall now consider a few standard cases for which such explicit solutions can be found.

2.1 Separable equations

Suppose the function $F(x, y)$ takes the special form $f(x)g(y)$, that is

$$\frac{dy}{dx} = f(x)g(y).$$

Then we may 'separate' x and y onto the respective sides of the equation to give

$$\frac{1}{g(y)} \, dy = f(x) \, dx,$$

so that

$$\int \frac{dy}{g(y)} = \int f(x) \, dx.$$

For any specific case we may integrate the two sides of the equation independently, and hence obtain a relationship between x and y. It will not normally be possible for this to be simplified to the point at which y is given as a specific function of x. It may even be necessary to carry out the respective quadratures by Simpson's rule or some similar numerical method. Let us now illustrate the idea with a few simple examples.

Example 2.1.1

$$\frac{dy}{dx} + x(1 - y^2)^{\frac{1}{2}} = 0.$$

Upon re-arranging the various terms, the equation becomes

$$\frac{dy}{(1 - y^2)^{\frac{1}{2}}} + x \, dx = 0.$$

Integration then yields

$$\sin^{-1}y + \tfrac{1}{2}x^2 = c.$$

An alternative form of this answer is

$$y = \sin (c - \tfrac{1}{2}x^2).$$

Example 2.1.2

$$y(1 + x^2)^{\frac{1}{2}} \frac{dy}{dx} = x(1 + y^2)^{\frac{1}{2}}.$$

This equation may be re-written as

$$\frac{y}{(1 + y^2)^{\frac{1}{2}}} \, dy = \frac{x}{(1 + x^2)^{\frac{1}{2}}} \, dx.$$

Integration leads to

$$(1 + y^2)^{\frac{1}{2}} = (1 + x^2)^{\frac{1}{2}} + c. \tag{2.2}$$

This is probably as far as we can go towards simplifying the answer in general. It is just possible that, having used the boundary condition to determine c, some simplification may follow in certain cases. For example, if the boundary condition is that $y = 0$ when $x = 0$, it then follows that $c = 0$. Thus equation (2.2) becomes

$$(1 + y^2)^{\frac{1}{2}} = (1 + x^2)^{\frac{1}{2}}.$$

Upon squaring both sides, then subtracting 1 from each side, we finally have

$$y = \pm x.$$

Example 2.1.3

$$\tan x \cos y \frac{dy}{dx} + 1 = 0.$$

After separating the variables this equation becomes

$$\cos y \, dy + \cot x \, dx = 0,$$

and this integrates to give

$$\sin y + \log (\sin x) = c.$$

Here again, no further simplification appears to be possible.

In addition to examples of the type we have already looked at, in which it is immediately obvious that the equation is separable, we may note that equations can sometimes be made separable by means of a suitable change of variable. The following examples will illustrate the possibilities.

Example 2.1.4

$$x\frac{\mathrm{d}y}{\mathrm{d}x} + y + x(1 - x^2y^2)^{\frac{1}{2}} = 0. \qquad (2.3)$$

If we recall that $x(\mathrm{d}y/\mathrm{d}x) + y$ is the derivative of the product $x.y$, it is clear that the change of variable

$$z = x.y$$

will effect a considerable simplification. Upon making this change of variable equation (2.3) becomes

$$\frac{\mathrm{d}z}{\mathrm{d}x} + x(1 - z^2)^{\frac{1}{2}} = 0,$$

which is simply Example 2.1.1 above. The solution is thus

$$z = \sin(c - \tfrac{1}{2}x^2),$$

or

$$y = \frac{\sin(c - \tfrac{1}{2}x^2)}{x}.$$

Example 2.1.5

$$(x + y)^2 \frac{\mathrm{d}y}{\mathrm{d}x} = 1.$$

In this instance we make the change of variable

$$z = x + y.$$

Thus

$$\frac{\mathrm{d}z}{\mathrm{d}x} = 1 + \frac{\mathrm{d}y}{\mathrm{d}x},$$

and the equation becomes

$$z^2\left\{\frac{\mathrm{d}z}{\mathrm{d}x} - 1\right\} = 1,$$

or

$$\frac{\mathrm{d}z}{\mathrm{d}x} = 1 + \frac{1}{z^2}.$$

It then follows that

$$dx = \frac{z^2}{1 + z^2}\, dz$$

$$= \left(1 - \frac{1}{1 + z^2}\right) dz,$$

and this now integrates to

$$x = z - \tan^{-1}z + c.$$

Transforming back to the original variable we have

$$x = x + y - \tan^{-1}(x + y) + c,$$

that is,

$$y = \tan^{-1}(x + y) - c.$$

Alternatively we may rewrite this result as

$$x = \tan(y + c) - y.$$

2.2 Homogeneous equations

We now consider the case when the function $F(x, y)$ in equation (2.1) takes the form $f(y/x)$. Consider, for example, the equation

$$x\frac{dy}{dx} = x + y.$$

Leaving to one side the term (dy/dx), we see that the terms involving the variables x and y are all of degree one. Thus we may divide by x and obtain

$$\frac{dy}{dx} = 1 + \frac{y}{x}.$$

Turning now to the most general homogeneous equation of degree and order one, we have

$$\frac{dy}{dx} = f\left(\frac{y}{x}\right).$$

It seems natural to take y/x as a new variable, so we write

$$y = v.x, \qquad (2.4)$$

whence

$$\frac{\mathrm{d}y}{\mathrm{d}x} = x\frac{\mathrm{d}v}{\mathrm{d}x} + v,$$

and so

$$x\frac{\mathrm{d}v}{\mathrm{d}x} + v = f(v).$$

This equation is separable, and may be written as

$$\frac{\mathrm{d}x}{x} = \frac{\mathrm{d}v}{f(v) - v},$$

so that

$$\log|x| = \int \frac{\mathrm{d}v}{f(v) - v}.$$

This relationship between v and x immediately leads, by using equation (2.4), to the required relationship between y and x.

Example 2.2.1

$$x\frac{\mathrm{d}y}{\mathrm{d}x} = x + y.$$

After division by x, the substitution $y = vx$ yields

$$x\frac{\mathrm{d}v}{\mathrm{d}x} + v = 1 + v.$$

The cancellation of the two terms in v leads to a particularly simple equation, which may be written as

$$\mathrm{d}v = \frac{\mathrm{d}x}{x},$$

with integral

$$v = \log|x| + c.$$

Thus, finally,

$$y = vx = x\{\log|x| + c\}.$$

Example 2.2.2

$$\frac{\mathrm{d}y}{\mathrm{d}x} = \frac{x + 3y}{3x + y}.$$

As before, the substitution

$$y = vx$$

leads to

$$x\frac{\mathrm{d}v}{\mathrm{d}x} + v = \frac{1 + 3v}{3 + v},$$

so

$$x\frac{\mathrm{d}v}{\mathrm{d}x} = \frac{1 + 3v}{3 + v} - v$$

$$= \frac{1 - v^2}{3 + v}.$$

Thus

$$\frac{\mathrm{d}x}{x} = \frac{3 + v}{1 - v^2} \, \mathrm{d}v$$

$$= \left(\frac{2}{1 - v} + \frac{1}{1 + v}\right) \mathrm{d}v.$$

Integration yields

$$\log |x| = -2 \log |1 - v| + \log |1 + v| + \text{const.},$$

and after taking exponentials this becomes

$$|x| = \frac{c \, |1 + v|}{(1 - v)^2},$$

or

$$|x| \left\{ 1 - \frac{y}{x} \right\}^2 = c \left| 1 + \frac{y}{x} \right|.$$

Upon multiplication by $|x|$, we have

$$(x - y)^2 = c \, |x + y|.$$

Sometimes a non-homogeneous equation can be converted into a homogeneous form by means of a suitable change of variable. Consider, for example, the equation

$$\frac{\mathrm{d}y}{\mathrm{d}x} = \frac{ax + by + c}{a'x + b'y + c'}.$$

Introduce new variables x_1, y_1, such that

$$x = x_1 + \alpha, \qquad y = y_1 + \beta.$$

Then

$$\frac{dy_1}{dx_1} = \frac{dy}{dx} = \frac{ax_1 + by_1 + (a\alpha + b\beta + c)}{a'x_1 + b'y_1 + (a'\alpha + b'\beta + c')}. \qquad (2.5)$$

Provided $ab' \neq a'b$, it is possible to choose α and β so that

$$a\alpha + b\beta + c = a'\alpha + b'\beta + c' = 0,$$

and equation (2.5) is then of homogeneous form.

Before discussing what happens when $ab' = a'b$, it is instructive to give a geometrical interpretation of the above change of variable. It is well known in co-ordinate geometry that $ax + by + c = 0$ represents a straight line, and that this straight line goes through the origin if $c = 0$. The above change of variable simply represents a change of origin, the new origin being at the point where the lines $ax + by + c = 0$ and $a'x + b'y + c' = 0$ cross. When $ab' = a'b$, the two lines are parallel, and the proposed new origin does not exist, hence the failure of the method.

The required change of variable in this case is as follows. We write

$$\frac{a}{a'} = \frac{b}{b'} = \lambda,$$

so the original equation may be written as

$$\frac{dy}{dx} = \frac{\lambda(a'x + b'y) + c}{a'x + b'y + c'}.$$

The introduction of the new variable

$$z = a'x + b'y$$

then leads to a separable equation.

Example 2.2.3

$$\frac{dy}{dx} = \frac{x + 3y + 1}{3x + y + 1}.$$

We write $x = x_1 + \alpha$, $y = y_1 + \beta$. Then, after substitution, it is readily found that

$$\frac{dy_1}{dx_1} = \frac{x_1 + 3y_1 + (\alpha + 3\beta + 1)}{3x_1 + y_1 + (3\alpha + \beta + 1)}.$$

By choosing $\alpha = \beta = -\frac{1}{4}$ this equation becomes precisely that solved as Example 2.2.2.

Example 2.2.4

$$(x + 2y)\frac{dy}{dx} = 2x + 4y + 1.$$

We note that x and y appear on both sides of the equation in the combination $x + 2y$. Accordingly we write

$$z = x + 2y,$$

whence it follows that

$$\frac{dz}{dx} = 1 + 2\frac{dy}{dx},$$

and so

$$\tfrac{1}{2}z\left(\frac{dz}{dx} - 1\right) = 2z + 1.$$

Therefore

$$\frac{dz}{dx} = \frac{5z + 2}{z},$$

and

$$dx = \frac{z}{5z + 2}\, dz.$$

Therefore

$$5x = \int \frac{5z}{5z + 2}\, dz,$$

$$= \int \left(1 - \frac{2}{5z + 2}\right) dz,$$

$$= z - \frac{2}{5}\log |5z + 2| + \text{const.}$$

We may re-arrange as

$$5(y - 2x) = \log|5x + 10y + 2| + c.$$

2.3 Linear equations

The most general form of a linear equation of order n is given by equation (1.4). For first-order equations, this may be written as

$$\frac{\mathrm{d}y}{\mathrm{d}x} + f(x)y = g(x). \tag{2.6}$$

The method of solution of this equation requires the use of what is known as an integrating factor. To see what this involves, we begin by examining the equation obtained by setting $g(x) = 0$. The resulting equation, though still linear of course, is also separable and may be integrated as such. Thus

$$\frac{\mathrm{d}y}{\mathrm{d}x} = -f(x)y,$$

and so

$$\frac{\mathrm{d}y}{y} = -f(x)\,\mathrm{d}x.$$

Therefore

$$\log y = -\int f(x)\,\mathrm{d}x + \text{const.}, \dagger$$

and

$$y = A \exp\{-\int f(x)\,\mathrm{d}x\}.$$

Let us write this equation in the form

$$y.\exp\{\int f(x)\,\mathrm{d}x\} = A,$$

and differentiate with respect to x. Then

$$\frac{\mathrm{d}y}{\mathrm{d}x}\cdot\exp\{\int f(x)\,\mathrm{d}x\} + y.\exp\{\int f(x)\,\mathrm{d}x\}.f(x) = 0,$$

or

$$\left\{\frac{\mathrm{d}y}{\mathrm{d}x} + f(x)y\right\}\exp\{\int f(x)\,\mathrm{d}x\} = 0.$$

† Strictly speaking, we should have $\log|y|$ on the left-hand side of this equation. If y is negative, the constant of integration may accordingly include an imaginary quantity, $\log(-1)$, that is, πi.

We see that this is exactly the equation from which we started, multiplied by

$$I(x) = \exp \{ \int f(x) \, dx \}.$$

This quantity is precisely that by which we must multiply $\left\{ \dfrac{dy}{dx} + f(x)y \right\}$ in order to obtain an exact derivative, which can then be integrated directly—for this reason $I(x)$ is referred to as an *integrating factor*.

Turning now to the original equation (2.6), we multiply through by $I(x)$. Thus

$$\left\{ \frac{dy}{dx} + f(x)y \right\} I = \frac{d}{dx} \{ yI(x) \} = g(x)I(x).$$

Therefore

$$yI(x) = \int g(x)I(x) \, dx + \text{const.}$$

and upon dividing by $I(x)$, we obtain y as a function of x.

Example 2.3.1

$$\frac{dy}{dx} + y = e^{-x}.$$

Here the integrating factor is

$$I(x) = \exp \{ \int 1 \, dx \} = e^x.$$

Thus

$$\frac{d}{dx}(y \, e^x) = 1;$$

therefore

$$y \, e^x = x + c,$$

and so

$$y = (x + c) \, e^{-x}.$$

Example 2.3.2

$$\frac{dy}{dx} + y = \sin x.$$

It will be noted that this example differs from the preceding one only in respect of the term on the right-hand side. Accordingly the integrating factor is unaltered, and it follows that

$$\frac{\mathrm{d}}{\mathrm{d}x}(y\,\mathrm{e}^x) = \mathrm{e}^x \sin x.$$

Therefore

$$y\,\mathrm{e}^x = \int \mathrm{e}^x \sin x\,\mathrm{d}x + c$$
$$= \tfrac{1}{2}\,\mathrm{e}^x\,(\sin x - \cos x) + c,$$

and so

$$y = \tfrac{1}{2}\,(\sin x - \cos x) + c\,\mathrm{e}^{-x}.$$

Example 2.3.3

$$(1 + x^2)\frac{\mathrm{d}y}{\mathrm{d}x} + xy = x(1 + x^2)^{\frac{1}{2}}.$$

Before we can determine the integrating factor for this linear equation we must first divide by $(1 + x^2)$, in order that the equation takes a form in which the coefficient of $\mathrm{d}y/\mathrm{d}x$ is unity. Thus

$$\frac{\mathrm{d}y}{\mathrm{d}x} + \frac{x}{1 + x^2}\,y = \frac{x}{(1 + x^2)^{\frac{1}{2}}},$$

and the integrating factor is

$$\begin{aligned}
I(x) &= \exp\left\{\int \frac{x}{1 + x^2}\,\mathrm{d}x\right\} \\
&= \exp\left\{\frac{1}{2}\int \frac{\mathrm{d}(1 + x^2)}{1 + x^2}\right\} \\
&= \exp\left\{\tfrac{1}{2}\log(1 + x^2)\right\} \\
&= (1 + x^2)^{\frac{1}{2}}.
\end{aligned}$$

Therefore

$$\frac{\mathrm{d}}{\mathrm{d}x}\{y(1 + x^2)^{\frac{1}{2}}\} = x.$$

Thus

$$y(1 + x^2)^{\frac{1}{2}} = \tfrac{1}{2}x^2 + c.$$

and

$$y = (\tfrac{1}{2}x^2 + c)(1 + x^2)^{-\frac{1}{2}}.$$

Example 2.3.4

$$y(x + y^2)\frac{dy}{dx} = 1.$$

The fact that this is a linear equation may well appear to be far from obvious. However, when it is rewritten as

$$\frac{dx}{dy} - yx = y^3, \qquad (2.7)$$

it is clearly a linear equation for x as a function of y. The student will realize from this example that it is unwise to have any pre-assigned notions as to which is the dependent and which the independent variable. To solve the linear equation (2.7), we note that the integrating factor is $\exp(-\tfrac{1}{2}y^2)$, and so

$$\frac{d}{dy}\{x \exp(-\tfrac{1}{2}y^2)\} = y^3 \exp(-\tfrac{1}{2}y^2)$$

and

$$\begin{aligned}
x \exp(-\tfrac{1}{2}y^2) &= \int y^3 \exp(-\tfrac{1}{2}y^2)\, dy + c \\
&= -(2 + y^2)\exp(-\tfrac{1}{2}y^2) + c.
\end{aligned}$$

Therefore

$$x = c \exp(\tfrac{1}{2}y^2) - y^2 - 2.$$

As with the earlier types of equation, there are certain equations which, with the help of a suitable change of variable, may be converted into a standard form. An important example is that of the so-called *Bernoulli* equation, which is of the form

$$\frac{dy}{dx} + P(x)y = Q(x)y^n.$$

This equation looks, superficially, very like a linear equation, the fly in the ointment being the solitary power of y at the very end of the equation. To solve, we divide out by the offending term, so that

$$\frac{1}{y^n}\frac{dy}{dx} + P(x)\frac{1}{y^{n-1}} = Q(x), \qquad (2.8)$$

and introduce a new variable

$$Y = \frac{1}{y^{n-1}}.$$

Since

$$\frac{dY}{dx} = -(n-1)\frac{1}{y^n}\frac{dy}{dx},$$

it follows that (2.8) becomes

$$-\frac{1}{n-1}\frac{dY}{dx} + P(x)Y = Q(x),$$

or

$$\frac{dY}{dx} - (n-1)P(x)Y = -(n-1)Q(x),$$

which is a linear equation for Y.

Example 2.3.5

$$\frac{dy}{dx} - xy = x^3 y^2.$$

We divide first by y^2. Therefore

$$\frac{1}{y^2}\frac{dy}{dx} - x.\frac{1}{y} = x^3.$$

We now write

$$Y = \frac{1}{y},$$

so that

$$\frac{dY}{dx} = -\frac{1}{y^2}\frac{dy}{dx},$$

and hence

$$\frac{dY}{dx} + xY = -x^3.$$

The integrating factor is $\exp\left(\frac{1}{2}x^2\right)$, so that

$$\frac{d}{dx}\left\{Y\exp\left(\tfrac{1}{2}x^2\right)\right\} = -x^3\exp\left(\tfrac{1}{2}x^2\right).$$

Therefore

$$Y \exp\left(\tfrac{1}{2}x^2\right) = c - \int x^3 \exp\left(\tfrac{1}{2}x^2\right) \mathrm{d}x$$
$$= c - (x^2 - 2) \exp\left(\tfrac{1}{2}x^2\right),$$

Therefore

$$Y = \frac{1}{y} = c \exp\left(-\tfrac{1}{2}x^2\right) - x^2 + 2.$$

2.4   Exact differentials

Consider the relationship

$$F(x, y) = \text{const.}$$

between x and y. Upon differentiating with respect to x we obtain

$$\frac{\partial F}{\partial x} + \frac{\partial F}{\partial y}\frac{\mathrm{d}y}{\mathrm{d}x} = 0. \tag{2.9}$$

Remembering that $(\partial F/\partial x)$ and $(\partial F/\partial y)$ are functions of x and y, it is clear that we have a first order, first degree differential equation.

Consider now the equation

$$P(x, y) + Q(x, y)\frac{\mathrm{d}y}{\mathrm{d}x} = 0. \tag{2.10}$$

Can we reverse the above procedure by finding a function $F(x, y)$ such that

$$\frac{\partial F}{\partial x} = P(x, y) \ and \ \frac{\partial F}{\partial y} = Q(x, y)? \tag{2.11}$$

Clearly we can always satisfy one of these conditions, but it will not always be possible to satisfy both. In those cases where both conditions can be satisfied the solution of equation (2.10) is obtained virtually by inspection.

Example 2.4.1

$$(x + 2y + 1) + (2x + y + 1)\frac{\mathrm{d}y}{\mathrm{d}x} = 0.$$

We note, in passing, that we could easily convert this equation into a homogeneous form. Alternatively, treating it as a possible case of an exact differential, we compare with equation (2.9), and ask whether the two conditions

$$\frac{\partial F}{\partial x} = x + 2y + 1, \tag{2.12}$$

$$\frac{\partial F}{\partial y} = 2x + y + 1, \tag{2.13}$$

can be satisfied simultaneously. Take equation (2.12) for example. Integration with respect to x, treating y as constant, yields

$$F = \tfrac{1}{2}x^2 + 2xy + x + f(y), \tag{2.14}$$

where the 'constant' of integration is the arbitrary function $f(y)$. Upon differentiation with respect to y, this yields

$$\frac{\partial F}{\partial y} = 2x + f'(y). \tag{2.15}$$

The results (2.13) and (2.15) are consistent if

$$f'(y) = y + 1,$$

that is, $\qquad f(y) = \tfrac{1}{2}y^2 + y + \text{const.},$

so equation (2.14) takes the form

$$F = \tfrac{1}{2}x^2 + 2xy + x + \tfrac{1}{2}y^2 + y + \text{const.} = \text{const.},$$

or

$$\tfrac{1}{2}x^2 + 2xy + x + \tfrac{1}{2}y^2 + y = c.$$

Example 2.4.2

$$(x + y + 1) + (2x + y + 1)\frac{dy}{dx} = 0.$$

Proceeding as above, we have

$$\frac{\partial F}{\partial x} = x + y + 1 \quad \text{and} \quad \frac{\partial F}{\partial y} = 2x + y + 1.$$

$$\downarrow$$

$$F = \tfrac{1}{2}x^2 + xy + x + f(y) \rightarrow \frac{\partial F}{\partial y} = x + f'(y).$$

The two values of $(\partial F/\partial y)$ are inconsistent, since no choice of the function $f(y)$ can enable the odd x to be accounted for. Accordingly this example is *not* an exact differential.

Example 2.4.3

$$(\cos x + \sec^2 x \tan y) + \tan x \sec^2 y \frac{\mathrm{d}y}{\mathrm{d}x} = 0.$$

This is of the form (2.9) if

$$\frac{\partial F}{\partial x} = \cos x + \sec^2 x \tan y \quad and \quad \frac{\partial F}{\partial y} = \tan x \sec^2 y.$$

$$\downarrow$$

$$F = \sin x + \tan x \tan y + f(y) \longrightarrow \frac{\partial F}{\partial y} = \tan x \sec^2 y + f'(y).$$

The two expressions for $(\partial F/\partial y)$ are consistent if

$$f'(y) = 0,$$

that is,

$$f(y) = \text{const.}$$

The solution is then

$$\sin x + \tan x \tan y = \text{const.}$$

The examples just considered are all characterized by the fact that the necessary integrations were easily carried out. In practice, of course, the integrations may be very much more complicated, with a considerable waste of effort if the example proves not to be an exact differential, to say nothing of the considerable scope for error. It is therefore helpful to have a simple way of deciding once and for all whether or not a given case is an exact differential.

Now equation (2.11) shows that if equation (2.10) is an exact differential then

$$\frac{\partial F}{\partial x} = P \quad \text{and} \quad \frac{\partial F}{\partial y} = Q.$$

Upon differentiating these results with respect to y and x, respectively, it follows that

$$\frac{\partial^2 F}{\partial y \partial x} = \frac{\partial P}{\partial y} \quad \text{and} \quad \frac{\partial^2 F}{\partial x \partial y} = \frac{\partial Q}{\partial x}.$$

Accordingly

$$\frac{\partial P}{\partial y} = \frac{\partial Q}{\partial x}, \tag{2.16}$$

provided that each of these quantities exists and is continuous. This condition (2.16) is, as shown above, a necessary condition. It may also be shown to be a sufficient condition, but we will not set out the proof here.

For Example 2.4.1 we have

$$P = x + 2y + 1 \quad \text{and} \quad Q = 2x + y + 1,$$

so

$$\frac{\partial P}{\partial y} = 2 \quad \text{and} \quad \frac{\partial Q}{\partial x} = 2,$$

showing that condition (2.16) is satisfied.

For Example 2.4.2 we have

$$P = x + y + 1 \quad \text{and} \quad Q = 2x + y + 1,$$

so

$$\frac{\partial P}{\partial y} = 1 \quad \text{and} \quad \frac{\partial Q}{\partial x} = 2,$$

showing that condition (2.16) is not satisfied.

Tutorial Examples

1. Integrate the following equations by separation of variables.

(a) $xy\dfrac{dy}{dx} = 1 + y^2$.

(b) $x(y + 2) + y(x + 2)\dfrac{dy}{dx} = 0$.

(c) $\dfrac{dr}{d\theta} = b\left(\cos\theta\dfrac{dr}{d\theta} + r\sin\theta\right)$.

A D E—B

2. By means of a suitable change of variable, convert the following equations to a separable form, and then integrate them.

(a) $\dfrac{dy}{dx} = \cos(x + y)$.

(b) $x + y\dfrac{dy}{dx} = x^2 + y^2$.

(c) $2x\dfrac{dy}{dx} = \dfrac{\sin x}{y} - y$.

3. Solve the following equations, for x as a function of t, subject to the prescribed conditions.

(a) $x' = 1 + x^2$, $x = 1$ when $t = 0$.

(b) $x' = 1 + t^2$, $x = 1$ when $t = 0$.

(c) $x' = xt$, $x = 1$ when $t = 2$.

(d) $x'' = -(x')^2$, $x = 2$ and $x' = 1$ when $t = 1$.

4. Solve the following homogeneous equations.

(a) $xy\dfrac{dy}{dx} + x^2 + y^2 = 0$.

(b) $x\dfrac{dy}{dx} + x + y = 0$.

(c) $\dfrac{dy}{dx} = \dfrac{2x + y}{2y - x}$.

(d) $x\dfrac{dy}{dx} + (x^2 + y^2)^{\frac{1}{2}} = y$.

5. Solve the following equations.

(a) $(3x - 5y)\dfrac{dy}{dx} = x - 3y + 2$.

(b) $(x + 2y - 5) - (3x + 6y + 7)\dfrac{dy}{dx} = 0$.

6. Solve the following linear equations.

   (a) $\dfrac{dy}{dx} + y \tanh x = \operatorname{sech}^3 x.$

   (b) $\operatorname{cosec} x \dfrac{dy}{dx} - y \sec x = \cot x.$

   (c) $(1 - x^2)\dfrac{dy}{dx} + 2xy = x - x^2.$

7. Solve the following exact differential equations.

   (a) $2x + 3y + 3(x + y^2)\dfrac{dy}{dx} = 0.$

   (b) $(2y + x)\dfrac{dy}{dx} + y = \cos x.$

8. Solve each of the following differential equations by *two* methods.

   (a) $(y - 2x)\dfrac{dy}{dx} = y.$

   (b) $(x + y + 1)\dfrac{dy}{dx} = x - y + 1.$

   (c) $y(2x + 3) + x(x + 3)\dfrac{dy}{dx} = 0.$

9. Solve the following miscellaneous equations.

   (a) $\dfrac{dy}{dx} = (y - x)^2.$

   (b) $x \cos y \dfrac{dy}{dx} - (1 + x^2) \sin y = 0.$

   (c) $(x - 2y - 3)\dfrac{dy}{dx} + 2 - 2x + 4y = 0.$

   (d) $x\dfrac{dy}{dx} + y = x^3 y^4.$

(e) $y\dfrac{\mathrm{d}y}{\mathrm{d}x} = y^2 \tanh x + x.$

(f) $(x^2 \operatorname{sech}^2 y + x \sinh y)\dfrac{\mathrm{d}y}{\mathrm{d}x} = \cosh y.$

10. Examine whether the differential equation

$$\cos^2 x + \sec x \tan y + \sin x \sec^2 y\,\frac{\mathrm{d}y}{\mathrm{d}x} = 0$$

has an integrating factor which is (a) a function of y alone, or (b) a function of x alone. Integrate the equation.

11. More generally you are to solve the equation

$$P(x, y)\,\mathrm{d}x + Q(x, y)\,\mathrm{d}y = 0.$$

Show that it has an integrating factor which depends on x alone provided that

$$\frac{1}{Q}\left\{\frac{\partial P}{\partial y} - \frac{\partial Q}{\partial x}\right\}$$

is a function of x alone.

CHAPTER 3

Other First-Order Equations

3.1 Equations from which x is missing

When x does not explicitly arise, the most general first-order equation is of the form

$$F(y, y') = 0.$$

If it is possible to express this result as

$$y' = f(y), \tag{3.1}$$

there is no problem, as this is a first-degree separable equation. If, on the other hand, the equation is more conveniently expressed as

$$y = g(y'), \tag{3.2}$$

and this cannot be readily inverted into the form (3.1), then we have an equation which is neither separable nor first-degree.

To solve equation (3.2) we first write $y' = p$ so that

$$y = g(p). \tag{3.3}$$

Upon differentiating with respect to x, this yields

$$p = g'(p)\frac{\mathrm{d}p}{\mathrm{d}x},$$

which is a separable equation in x and p. Thus

$$x = \int g'(p)\frac{\mathrm{d}p}{p}. \tag{3.4}$$

Equations (3.3) and (3.4) give y and x in terms of $p\,(=y')$, which may be regarded as a parameter.

Example 3.1.1

$$y = p^2 + 2p^3.$$

Differentiate with respect to x. Therefore

$$p = (2p + 6p^2)\frac{dp}{dx},$$

and so

$$x = \int (2 + 6p)\, dp†$$
$$= 2p + 3p^2 + c.$$

Thus x and y are both known in terms of the parameter p.

3.2 Clairaut's equation

Clairaut's equation is of the form

$$y = xy' + f(y')$$
$$xp + f(p). \tag{3.5}$$

As before we find a solution by first differentiating with respect to x. This yields

$$p = x\frac{dp}{dx} + p + f'(p)\frac{dp}{dx},$$

or

$$\frac{dp}{dx}\{x + f'(p)\} = 0. \tag{3.6}$$

There are now two possibilities. The first possibility is that

$$\frac{dp}{dx} = 0,$$

so that

$$p = \text{const.} = a,$$

and hence

$$y = ax + b. \tag{3.7}$$

---

† We cancel by p, so excluding the possibility $p = 0$. By virtue of the equation, $y = p^2 + 2p^3$, this leads only to the trivial solution $y = 0$.

There is clearly something of an anomaly here, in that we began with the first-order equation (3.5), and yet apparently have obtained a solution with two arbitrary constants. The second constant has arisen because we differentiated equation (3.5), and the result (3.7) is not a solution of equation (3.5) for all values of a and b. In fact, by substitution, it is seen that equation (3.7) represents a solution of equation (3.5) provided that

$$ax + b = ax + f(a),$$

that is

$$b = f(a),$$

so the solution of equation (3.5) then becomes

$$y = ax + f(a). \tag{3.8}$$

We return now to the second possible solution of equation (3.6), namely

$$x + f'(p) = 0. \tag{3.9}$$

This equation, taken in conjunction with the original equation (3.5), provides a parametrically defined relationship between x and y.

It is not difficult to show that the solution given by equations (3.5) and (3.9) is the envelope of the straight lines given by the solution of equation (3.8). To find this envelope we need to calculate where two adjacent lines cross, and this is done by solving simultaneously equation (3.8) with the equation obtained by differentiating it with respect to a, that is, with

$$0 = x + f'(a). \tag{3.10}$$

We see that equations (3.8) and (3.10) are indeed identical with equations (3.5) and (3.9), once it is remembered that a and p, respectively, are to be regarded as parameters.

Example 3.2.1

$$xp^2 - yp + 1 = 0.$$

This equation may be written in Clairaut's form as

$$y = xp + \frac{1}{p}. \tag{3.11}$$

Differentiating with respect to x, we find that

$$p = x\frac{\mathrm{d}p}{\mathrm{d}x} + p - \frac{1}{p^2}\frac{\mathrm{d}p}{\mathrm{d}x},$$

and hence

$$\frac{\mathrm{d}p}{\mathrm{d}x}\left(x - \frac{1}{p^2}\right) = 0.$$

Either

$$\frac{\mathrm{d}p}{\mathrm{d}x} = 0,$$

that is,

$$p = \text{const.} = a,$$

and hence,

$$y = ax + b$$
$$= ax + \frac{1}{a}, \tag{3.12}$$

for the reasons given above.

Alternatively,

$$x = \frac{1}{p^2}, \tag{3.13}$$

and eliminating p between equations (3.11) and (3.13) we have

$$y = x(\pm x^{-\frac{1}{2}}) \pm x^{\frac{1}{2}}$$
$$= \pm 2x^{\frac{1}{2}},$$

or

$$y^2 = 4x. \tag{3.14}$$

The straight lines (3.12) are, of course, the tangents to the parabola (3.14).

3.3   The generalized Clairaut equation

Consider now the equation

$$y = xf(y') + g(y')$$
$$= xf(p) + g(p).$$

Differentiation with respect to x yields

$$p = xf'(p)\frac{dp}{dx} + f(p) + g'(p)\frac{dp}{dx},$$

or

$$p - f(p) = \frac{dp}{dx}\{xf'(p) + g'(p)\}. \tag{3.15}$$

If $f(p) = p$ the equation is, of course, Clairaut's equation and we proceed as discussed earlier. In the more general case when $f(p) \neq p$, we may rewrite equation (3.15) as

$$\frac{dx}{dp} + \frac{f'(p)}{f(p) - p}x = -\frac{g'(p)}{f(p) - p},$$

which is a linear equation for x in terms of p. The solution of this equation, when taken in conjunction with the original equation, gives x and y in terms of the parameter p.

Example 3.3.1

$$y = xp^2 + p. \tag{3.16}$$

Differentiation yields

$$p = p^2 + 2xp\frac{dp}{dx} + \frac{dp}{dx},$$

and hence

$$(p - p^2)\frac{dx}{dp} - 2px = 1.$$

Therefore

$$\frac{dx}{dp} + \frac{2}{p - 1}x = \frac{1}{p(1 - p)}.$$

The integrating factor is simply $(p - 1)^2$, and hence

$$\frac{d}{dp}\{x(p - 1)^2\} = \frac{1}{p} - 1.$$

Integration yields

$$x(p - 1)^2 = \log|p| - p + c. \tag{3.17}$$

For any chosen value of p, equation (3.17) yields x, and the corresponding value of y follows from equation (3.16).

3.4 The Riccati equation

The equation
$$y' = A(x)y^2 + B(x)y + C(x) \qquad (3.18)$$
is called a Riccati equation. If $A \equiv 0$ the equation degenerates to linear form. Likewise, if $C \equiv 0$ it becomes a Bernoulli equation. If neither A nor C are identically zero, the equation can in general be solved only by numerical methods. However, as we shall now show, if any solution $y_1(x)$ can be found, the general solution can be obtained by means of the substitution
$$y = y_1 + \frac{1}{v}.$$

With this substitution, equation (3.18) becomes
$$y_1' - \frac{1}{v^2}v' = A\left(y_1^2 + \frac{2y_1}{v} + \frac{1}{v^2}\right) + B\left(y_1 + \frac{1}{v}\right) + C$$

But
$$y_1' = Ay_1^2 + By_1 + C.$$

and after subtraction it follows that
$$-\frac{1}{v^2}v' = (2Ay_1 + B).\frac{1}{v} + A.\frac{1}{v^2},$$

or
$$-v' = (2Ay_1 + B)v + A.$$

This is a linear equation for v, which is easily solved.

Example 3.4.1
$$y = xy^2 - 2x^2y + 2.$$

By inspection, this equation has a solution $y = 2x$. We accordingly write
$$y = 2x + \frac{1}{v}. \qquad (3.19)$$

After the appropriate algebra, which we omit, it transpires that

$$v' + 2x^2 v = -x.$$

The integrating factor is

$$\exp\left(\tfrac{2}{3}x^3\right),$$

whence

$$\frac{\mathrm{d}}{\mathrm{d}x}\{v \exp\left(\tfrac{2}{3}x^3\right)\} = x(\exp \tfrac{2}{3}x^3),$$

and so

$$v = A \exp\left(-\tfrac{2}{3}x^3\right) - \exp\left(-\tfrac{2}{3}x^3\right) \int^x x \exp\left(\tfrac{2}{3}x^3\right) \mathrm{d}x.$$

The original variate y then follows from equation (3.19).

Tutorial Examples

Obtain the general solution and singular solutions, if any, of each of the following equations.

1. $y = y' + \tfrac{1}{2}(y')^2.$ 3. $y - x = (y')^2(1 - \tfrac{2}{3}y').$

2. $(y - xy')^2 = 1 + (y')^2.$ 4. $x^2 y' = x(y - 1) + (y - 1)^2.$

CHAPTER 4

Applications of First-Order Equations

Our main aim in this chapter is to try and understand the way in which differential equations arise in practical situations. We shall do this by examining a variety of problems for which the resulting equations are sufficiently elementary to be within the compass of the average student. Needless to say, we shall need to make simplifications and approximations and the reader should accept that the detailed justification for some of these steps is beyond the scope of this book. The ability to make approximations which lead to a simplification of the mathematics without loss of realism is something which can only come with experience, and the examples chosen will illustrate something of what can be achieved.

4.1 Particle projected vertically under negligible air resistance

We measure x vertically *upwards* from the point of projection. The motion of the particle is described by Newton's law, which states that the rate of change of the momentum of the particle will equal the force exerted on it. Now if the mass of the particle is m, then the upward momentum will be mv, where $v = (dx/dt)$ is the upward velocity. In the absence (or neglect!) of air resistance the only force acting is gravity; this force equals mg in magnitude, and acts *downwards*, that is,

36

$-mg$ upwards. Thus the required equation is

$$\frac{\mathrm{d}}{\mathrm{d}t}(mv) = -mg,$$

or

$$\frac{\mathrm{d}v}{\mathrm{d}t} = -g,$$

since the mass m is constant.

This elementary equation easily integrates to give

$$v = A - gt.$$

If v_0 is the velocity with which the particle is projected, then $v = v_0$ when $t = 0$, and so

$$v = v_0 - gt. \tag{4.1}$$

This equation shows how the *velocity* of the particle varies with time. To determine its *position* we write

$$\frac{\mathrm{d}x}{\mathrm{d}t} = v = v_0 - gt,$$

which integrates to give

$$x = B + v_0 t - \tfrac{1}{2}gt^2,$$

that is,

$$x = v_0 t - \tfrac{1}{2}gt^2, \tag{4.2}$$

since $x = 0$ when $t = 0$.

Any required property of the motion may then be determined from equations (4.1) and (4.2). For example, the particle reaches its maximum height when $v = 0$; equation (4.1) shows that this happens at a time

$$t = \frac{v_0}{g},$$

and equation (4.2) then shows that this maximum height is

$$v_0 \cdot \frac{v_0}{g} - \tfrac{1}{2}g\left(\frac{v_0}{g}\right)^2 = \frac{v_0^2}{2g}.$$

Again, the particle returns to its original position, $x = 0$, when

$$v_0 t - \tfrac{1}{2}gt^2 = 0,$$

that is,

$$t = 0 \quad \text{or} \quad \frac{2v_0}{g}.$$

Since the time of ascent is v_0/g, the time of descent is also v_0/g. The velocity of the particle on return to the point of projection is obtained by setting $t = 2v_0/g$ in equation (4.1), and is

$$v_0 - g\left(\frac{2v_0}{g}\right) = -v_0.$$

Thus the velocity is downwards (predictably!) and of the same magnitude as the velocity of projection.

4.2 Particle projected vertically with resistance proportional to velocity

Suppose, in order to make the problem just considered somewhat more realistic, we assume that the particle is subject to air resistance of magnitude $mk\,|v|$. Then the requisite form of Newton's law is

$$\frac{\mathrm{d}}{\mathrm{d}t}(mv) = \text{net upward force acting on the particle}$$

$$= -mg - mkv,$$

the sign of the last term being negative since the resistance acts downwards when the particle moves upwards. Thus

$$\frac{\mathrm{d}v}{\mathrm{d}t} = -g - kv.$$

This equation is linear. (It is also separable, incidentally!) The integrating factor is e^{kt}. Therefore

$$\frac{d}{dt}(v\,e^{kt}) = -g\,e^{kt},$$

and hence

$$v\,e^{kt} = A - \frac{g}{k}e^{kt}$$

$$= \left(v_0 + \frac{g}{k}\right) - \frac{g}{k}e^{kt},$$

since $v = v_0$ when $t = 0$. Thus, finally,

$$v = \left(v_0 + \frac{g}{k}\right) e^{-kt} - \frac{g}{k}. \tag{4.3}$$

We may determine x by integration of equation (4.3), and after applying the boundary condition that $x = 0$ when $t = 0$ we find that

$$x = \frac{1}{k}\left(v_0 + \frac{g}{k}\right)(1 - e^{-kt}) - \frac{g}{k}t. \tag{4.4}$$

The various properties of the solution follow as before. The maximum height is reached when $v = 0$, that is, when

$$\left(v_0 + \frac{g}{k}\right) e^{-kt} = \frac{g}{k}.$$

This height is then given by equation (4.4) as

$$x = \frac{1}{k}\left(v_0 + \frac{g}{k}\right) - \frac{g}{k^2} - \frac{g}{k^2} \log\left(1 + \frac{kv_0}{g}\right)$$

$$= \frac{v_0}{k} - \frac{g}{k^2} \log\left(1 + \frac{kv_0}{g}\right). \tag{4.5}$$

If the resistance is small, that is, k small, we may write

$$\log\left(1 + \frac{kv_0}{g}\right) = \frac{kv_0}{g} - \tfrac{1}{2}\left(\frac{kv_0}{g}\right)^2 + \tfrac{1}{3}\left(\frac{kv_0}{g}\right)^3 + \cdots,$$

so (4.5) shows that

$$x_{\max} = \frac{v_0^2}{2g}\left(1 - \frac{2kv_0}{3g} + \cdots\right)$$

$$< \frac{v_0^2}{2g};$$

in other words, the height reached by the particle is reduced by the effects of the air resistance.

If we wish to relate v and x we can either eliminate t between equations (4.3) and (4.4), or we can proceed directly as follows.

$$\frac{dv}{dt} = \frac{dv}{dx} \cdot \frac{dx}{dt} = v\frac{dv}{dx} = -g - kv,$$

which is separable. Thus

$$-dx = \frac{v}{g + kv}\, dv,$$

and so

$$
\begin{aligned}
A - kx &= \int \frac{kv\, dv}{g + kv} \\
&= \int \left(1 - \frac{g}{g + kv}\right) dv \\
&= v - \frac{g}{k} \log \left|v + \frac{g}{k}\right|.
\end{aligned}
\tag{4.6}
$$

Before carrying the solution any further, we may note that $v + g/k$ is always positive, for the following reasons. As the particle falls, its speed will tend to increase (that is, v becomes more negative) under the influence of gravity, but this tendency will be opposed by the air resistance. There is a limiting speed which the particle cannot exceed, determined by the condition that air resistance and gravity exactly balance, that is, $g + kv = 0$, that is, $v = -g/k$. This conclusion is borne out by equation (4.3) which shows that $v + g/k$ is positive but tends to zero as $t \to \infty$.

On this basis we may ignore the modulus sign in equation (4.6), which becomes

$$A - kx = v - \frac{g}{k} \log \left(v + \frac{g}{k}\right).$$

The condition that $v = v_0$ when $x = 0$ determines A, and this gives

$$A = v_0 - \frac{g}{k} \log \left(v_0 + \frac{g}{k}\right),$$

so that

$$kx = \frac{g}{k} \log \frac{v + g/k}{v_0 + g/k} + v_0 - v. \qquad (4.7)$$

When the particle passes the point of projection (on the way down!) the velocity is given by putting $x = 0$ in (4.7). Thus

$$\frac{g}{k} \log \frac{v + g/k}{v_0 + g/k} - v + v_0 = 0.$$

If we write $kv/g = V$ and $kv_0/g = V_0$, this becomes

$$\log \frac{V + 1}{V_0 + 1} - V + V_0 = 0,$$

or

$$f(V) = V - \log(1 + V) = V_0 - \log(1 + V_0). \qquad (4.8)$$

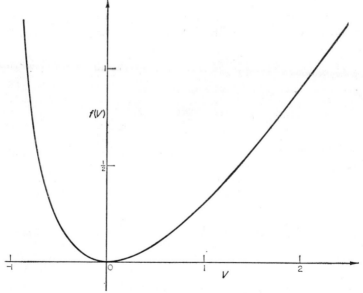

FIG. 4.1   The function $f(V) = V - \log(1 + V)$

The graph of $f(V)$ is as shown in Fig. 4.1. For any given positive value of V_0 it is seen that the second solution of equation (4.8) is negative, and that it is smaller in magnitude than V_0. Thus the speed is *less* than the speed of projection.

4.3 Particle escaping from earth's gravitational attraction

Let us now consider a particle projected vertically from the earth, assumed to be a sphere of radius R, with sufficient speed that it rises to a height comparable with R. Now the gravitational force is proportional to r^{-2}, where r is measured from the centre of the earth, and can no longer be assumed to be constant when r varies by an amount of the magnitude now considered. Thus the 'downward' pull of gravity on a particle of mass m is of magnitude.

$$mg\left(\frac{R}{r}\right)^2,$$

where g is the gravitation in the vicinity of the earth's surface. Ignoring air resistance, the motion of the particle must satisfy

$$\frac{d}{dt}(mv) = -mg\left(\frac{R}{r}\right)^2,$$

or

$$\frac{dv}{dt} = v\frac{dv}{dr} = -g\left(\frac{R}{r}\right)^2.$$

Taking the second form for the acceleration, $v(dv/dr)$, integration then yields

$$\tfrac{1}{2}v^2 = A + \frac{gR^2}{r}.$$

But $v = v_0$ when $r = R$, so

$$A = \tfrac{1}{2}v_0^2 - gR,$$

and

$$\tfrac{1}{2}v^2 = \tfrac{1}{2}v_0^2 - gR + \frac{gR^2}{r}.$$

If

$$\tfrac{1}{2}v_0^2 - gR > 0,$$

then the value of $\frac{1}{2}v^2$, which diminishes as r increases, will never diminish to zero, and so the particle will ultimately escape from the earth's gravitational pull. The minimum projection velocity which will achieve this is called the velocity of escape, and is thus equal to

$$v_0 = (2gR)^{\frac{1}{2}}.$$

4.4 Cooling of a heated body

Consider a body, initially heated to a temperature T_0, placed in an atmosphere of ambient temperature T_a. It is reasonable to suppose that, at any instant at which the mean temperature of the body is T, the rate $-dT/dt$ at which its temperature is falling, is roughly proportional to the excess $T - T_a$ of body temperature over ambient temperature. Thus

$$-\frac{dT}{dt} = k(T - T_a). \tag{4.9}$$

This law, known as Newton's law of cooling, has been verified experimentally over a surprisingly wide range of conditions. We note that equation (4.9) is separable, so that

$$\frac{dT}{T - T_a} = -k \, dt,$$

and

$$\log(T - T_a) = -kt + \text{const.}$$

Upon taking exponentials this becomes

$$\begin{aligned} T - T_a &= A \, e^{-kt} \\ &= (T_0 - T_a) \, e^{-kt}. \end{aligned} \tag{4.10}$$

by virtue of the initial condition that $T = T_0$ when $t = 0$. We note from equation (4.10) that the temperature differential diminishes exponentially with time.

4.5 A simple electrical circuit

A simple electrical circuit consists of a capacitance, an inductance, and a resistance, together with an applied electromotive force (e.m.f.). In general the equation governing such a circuit is of second order, but when the capacitance is zero, the equation is of first order.

A resistance does what its name implies—it resists the passage of current through a circuit. If the resistance is R and the current I, then the drop in voltage (or potential) across the resistance is RI. An inductance resists *change* of current, and helps to keep things steady. The drop in voltage across an inductance L is $L(dI/dt)$.

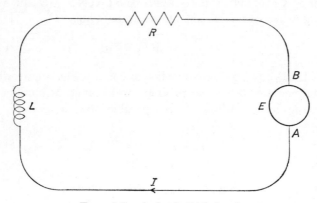

FIG. 4.2   A simple L/R circuit

Thus in a circuit such as that shown in Fig. 4.2, the potential drop between A and B as the current flows as shown is

$$L\frac{dI}{dt} + RI,$$

and this must exactly balance the potential gain at the battery, say $E(t)$. Hence

$$L\frac{dI}{dt} + RI = E(t).$$

This is a simple linear equation, with an integrating factor $e^{Rt/L}$.

4.6 Flow of liquid from a tank

Consider the problem of liquid flowing out of a tank through a small opening at the base. It is a matter of experimental observation that, when the liquid flows out freely under the action of the head of liquid above it, the speed at which it flows is proportional to $\sqrt{(gy)}$, where y is the height of the liquid above the opening. Thus the volume of liquid flowing out per unit time through a hole of area A will be $kA\sqrt{(gy)}$. The constant of proportionality k depends upon the details of the shapes of the opening and of the tank as a whole, but it is usually of order unity.

Suppose that $S(y)$ is the cross-sectional area of the tank at a height y. In time dt the level of the liquid falls by an amount $-dy$. Then the volume of liquid leaving the tank is equal to $kA\sqrt{(gy)}\,dt$; it is also equal to $-S(y)\,dy$, since this is the volume 'vacated' as the level falls. Thus

$$kA\sqrt{(gy)} = -S(y)\frac{dy}{dt}, \qquad (4.11)$$

which is a separable equation relating y and t.

As an illustration, consider liquid contained within a funnel of semi-angle α, as shown in Fig. 4.3. The funnel is initially filled to a height h and the liquid flows out through a tube of radius a. Here the cross-section of the funnel is a circle of radius $y \tan \alpha$, so $S(y)$ is equal simply to $\pi(y \tan \alpha)^2$. Likewise A is equal to πa^2. With these values, equation (4.11) becomes

$$k\pi a^2 \sqrt{(gy)} = -\pi(y \tan \alpha)^2 \frac{dy}{dt},$$

or

$$\frac{ka^2 \sqrt{g}}{\tan^2 \alpha}\, dt = -y^{3/2}\, dy.$$

Integration yields

$$\frac{ka^2\sqrt{g}}{\tan^2 \alpha}t = -\tfrac{2}{5}y^{5/2} + \text{const.}$$

$$= \tfrac{2}{5}(h^{5/2} - y^{5/2}), \qquad (4.12)$$

since $y = h$ when $t = 0$. Equation (4.12) is the required relationship between y and t.

The tank is empty when $y = 0$, and this will occur after a time

$$t = \frac{2}{5k}h^{5/2}\frac{\tan^2 \alpha}{a^2 g^{\frac{1}{2}}}. \qquad (4.13)$$

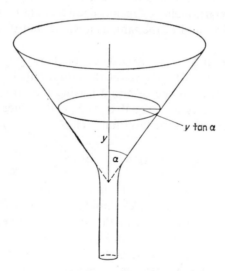

FIG. 4.3 Funnel of semi-angle α

The reasons for the various factors which arise in equation (4.13) are as follows. The volume of liquid originally in the funnel is proportional to $h(h \tan \alpha)^2$. The time to empty the funnel will be proportional to this; it will be inversely proportional to the size of the opening, that is, to a^2, and to the mean outflow velocity, that is, to $\sqrt{(gh)}$. The appropriate ratio of these quantities yields

$$\frac{h^3 \tan^2 \alpha}{a^2 \sqrt{(gh)}},$$

which is precisely as shown in equation (4.13).

4.7 Formation of a chemical compound

Consider two elements X, Y, which combine in such a way that a units of X and b units of Y form one unit of Z. Initially, at time $t = 0$, we assume that there are x_0 units of X, y_0 units of Y, but no units of Z. At time $t > 0$ there are z units of Z, and, accordingly, $(x_0 - az)$ units of X plus $(y_0 - bz)$ units of Y remain uncompounded. It is reasonable to suppose, as a very rough guide, that

$$\frac{dz}{dt} = k(x_0 - az)(y_0 - bz), \qquad (4.14)$$

since dz/dt must be zero when the free amount of either element falls to zero. Now equation (4.14) is separable, and may be written as

$$k\,dt = \frac{dz}{(az - x_0)(bz - y_0)}$$
$$= \frac{dz}{bx_0 - ay_0} \left\{ \frac{a}{ax - x_0} - \frac{b}{bz - y_0} \right\},$$

provided that $bx_0 - ay_0$ is non-zero. In such circumstances the equation integrates to give

$$k(bx_0 - ay_0)t - \log \frac{az - x_0}{bz - y_0} + \text{const.}$$

Use of the initial condition, $z = 0$ when $t = 0$, leads to

$$k(bx_0 - ay_0)t = \log \frac{1 - \dfrac{az}{x_0}}{1 - \dfrac{bz}{y_0}},$$

or

$$\frac{1 - \dfrac{az}{x_0}}{1 - \dfrac{bz}{y_0}} = \exp\{k(bx_0 - ay_0)t\}.$$

Thus, if $bx_0 - ay_0 > 0$ then $z \rightarrow (y_0/b)$ as $t \rightarrow \infty$. Likewise, if $bx_0 - ay_0 < 0$ then $z \rightarrow (x_0/a)$ as $t \rightarrow \infty$. In each case the limiting value of z is, of course, determined by which element is used up first in making the compound.

When $bx_0 - ay_0 = 0$, equation (4.14) may be written as

$$\frac{\mathrm{d}z}{\mathrm{d}t} = kab(\lambda - z)^2,$$

where

$$\lambda = \frac{x_0}{a} = \frac{y_0}{b}.$$

This separable equation integrates directly, so that

$$\begin{aligned}
kabt &= \int \frac{\mathrm{d}z}{(\lambda - z)^2} + \text{const.}\\
&= \frac{1}{\lambda - z} + A\\
&= \frac{1}{\lambda - z} - \frac{1}{\lambda} \text{ (by initial conditions)}\\
&= \frac{z}{\lambda(\lambda - z)}.
\end{aligned}$$

Hence

$$kbx_0 t = \frac{z}{\lambda - z},$$

and

$$z = \frac{kx_0 y_0 t}{1 + kbx_0 t}. \tag{4.15}$$

As $t \rightarrow \infty$, $z \rightarrow x_0/a(y_0 b)$, and the approach to the limit takes place algebraically rather than exponentially.

Tutorial Examples

1. A simple electrical circuit consists of an inductance L, a resistance R, and a constant applied e.m.f. equal to E_0. The circuit is linked at $t = 0$. Show that the current I is given by

$$I = \frac{E_0}{R}\left\{1 - \exp\left(-\frac{Rt}{L}\right)\right\}.$$

2. The rate of decay of a radioactive substance may be assumed to be proportional to the amount remaining. Show that the amount present will decay exponentially.

3. A bath is full of cold water at temperature T_C. Hot water, of temperature T_H, runs in at such a rate as would fill an empty bath in time t_0. Assuming that mixing occurs instantly, so that the water running out is taken to be at the mean temperature T, show that the mean temperature after a time t_0 is

$$T = T_H + \frac{T_C - T_H}{e}.$$

4. A particle of mass m is projected vertically upwards from the ground with initial velocity v_0, and is subject to air resistance of magnitude mkv^2, where v denotes the velocity at any time t. Find the maximum height x_m reached by the particle, and its speed v_1 on return to the ground.

5. The mass m of a raindrop increases with time as it falls through a cloud. Assuming that

$$m = m_0 \exp\left(t/t_0\right),$$

where m_0 is the mass at time $t = 0$, and t_0 is the time taken for the mass to increase by a factor e, and ignoring the air resistance, derive an equation representing the downward motion of the raindrop.

If the drop is approximately at rest when $t = 0$, show that its downward velocity is subsequently

$$v = gt_0\{1 - \exp(-t/t_0)\}.$$

Is this consistent with the result that $v = gt$ for a drop of constant mass?

6. A hemispherical tank of radius r is full of water. At time $t = 0$ a tap is opened, and water drains through a circular hole of radius a at the base of the tank. Assume that, when the height of the liquid in the tank is y, the volume of water which runs through the hole in unit time is

$$k(\pi a^2)\sqrt{(gy)}.$$

Show that the time taken for the tank to empty is

$$\frac{14}{15k} \frac{r^{5/2}}{a^2\sqrt{g}}.$$

7. A spherical mothball loses mass by evaporation at a rate which is proportional to its surface area. If it loses half its mass in 75 days, show that it will take approximately 1 year to disappear completely.

8. As a raindrop falls, its mass increases at a rate proportional to its surface area. At time $t = 0$ the radius is r_0, and this doubles in a time t_0. Ignoring air resistance, and assuming that the drop is at rest when $t = 0$, show that the downward velocity at time t is

$$v = \tfrac{1}{4}gt_0\left\{1 + \frac{t}{t_0} - \left(1 + \frac{t}{t_0}\right)^{-3}\right\}.$$

9. When a rope is wound tightly around a rough cylinder, it is observed that the change in tension per unit length is proportional to the tension. The constant of proportionality is μ/a, where μ is the coefficient of friction between the rope and the cylinder, and a is the radius of the cylinder. Show that when the rope is wound around the cylinder n

times, a man holding one end can resist a force which is greater than he can exert by a factor $\exp(2\pi\mu n)$.

10. A particle of mass m is projected vertically upwards with speed v_0. It experiences an air resistance equal to $mkv^2 f(r)$, where $f(r)$ is equal to 1 near the surface of the earth and diminishes to zero at points outside the earth's atmosphere. The gravitational attraction is inversely proportional to the square of the distance from the centre of the earth. Show that the particle will escape from the earth's atmosphere if

$$v_0^2 \geqslant 2gr_0^2 \int_{r_0}^{\infty} \exp\left\{2k \int_r^r f(r)\, dr\right\} \frac{dr}{r^2},$$

where r_0 (assumed constant) is the radius of the earth.

CHAPTER 5

Second-Order Equations with Constant Coefficients

The most general second-order linear ordinary differential equation may be written as

$$a(x)y'' + b(x)y' + c(x)y = f(x).$$

For the case of constant coefficients we may write this more simply as

$$ay'' + by' + cy = f(x). \tag{5.1}$$

Basically the general solution of this equation may be expressed, as was shown in Section 1.4, as the sum of the complementary function and a particular integral. The complementary function is the general solution of the equation

$$ay'' + by' + cy = 0, \tag{5.2}$$

and we shall begin by discussing how this equation may be solved.

5.1 Calculation of the complementary function

It is very tempting simply to state that a solution of equation (5.2) can be obtained in the form of an exponential, and to be content that it works! But it is more enlightening if we can get some indication as to why this type of solution might be anticipated in advance. Possibly some help can be found if we look at special cases.

Let us see first of all what happens to equation (5.2) when $a = 0$. The equation then becomes

$$by' + cy = 0.$$

This is now a first-order equation, which is easily solved, being both linear and separable. The solution is

$$y = A \exp\left(-\frac{c}{b}x\right). \tag{5.3}$$

Alternatively, what happens when $c = 0$? The equation is then

$$ay'' + by' = 0,$$

which can be regarded as a first-order equation for y'. The solution is

$$y' = A \exp\left(-\frac{b}{a}x\right),$$

and so

$$y = A_1 \exp\left(-\frac{b}{a}x\right) + B. \tag{5.4}$$

Again, when $b = 0$, the equation is

$$ay'' + cy = 0.$$

After multiplication by $2y'$ this becomes

$$2ay'y'' + 2cyy' = 0,$$

which integrates to

$$a(y')^2 + cy^2 = A.$$

Depending upon whether a and c have the same or opposite signs, this may be written in the form

$$y' = \alpha(y_0^2 - y^2)^{\frac{1}{2}} \quad \text{or} \quad y' = \alpha(y_0^2 + y^2)^{\frac{1}{2}},$$

which integrate to give

$$y = y_0 \sin(\alpha x + B) \quad \text{or} \quad y = y_0 \sinh(\alpha x + B). \tag{5.5}$$

By examining equations (5.3)–(5.5), it is seen that in each case the solution takes the form of exponentials, remembering that the sine function in equation (5.5) is merely an exponential of imaginary argument. We are accordingly led to investigate the possibility that the equation (5.2) has exponential type solutions in general, and this we shall now do.

We look for a solution of equation (5.2) of the form

$$y = A e^{\lambda x}$$

for some values of A and λ. It follows that
$$y' = A\lambda\,e^{\lambda x} \quad \text{and} \quad y'' = A\lambda^2\,e^{\lambda x},$$
so upon substitution into equation (5.2) it follows that
$$(a\lambda^2 + b\lambda + c)A\,e^{\lambda x} = 0.$$
There is a non-trivial solution, for which $y = A\,e^{\lambda x}$ is not identically zero, if λ is chosen to satisfy
$$a\lambda^2 + b\lambda + c = 0.$$
This equation has two roots, which are real and different if $b^2 > 4ac$, real and equal if $b^2 = 4ac$, complex if $b^2 < 4ac$.

Case 1. Real (unequal) roots. $\lambda = \lambda_1$ or λ_2.

Then $y = A\exp(\lambda_1 x)$ and $y = B\exp(\lambda_2 x)$ are both solutions of the equation for *any* values of A and B. But the equation is linear, so we can superimpose solutions. Thus
$$y = A\exp(\lambda_1 x) + B\exp(\lambda_2 x) \tag{5.6}$$
is also a solution. But it has two arbitrary constants; it is therefore the general solution.

Example 5.1.1
$$y'' + 3y' + 2y = 0.$$
Try $\quad y \propto e^{\lambda x} \;\longrightarrow\; \lambda^2 + 3\lambda + 2 = 0.$
That is, $\quad (\lambda + 1)(\lambda + 2) = 0.$
Therefore $\quad \lambda = -1$ or -2,
Therefore
$$y = A\,e^{-x} + B\,e^{-2x}.$$

Case 2. Equal roots (that is, $b^2 = 4ac$).

The equation $a\lambda^2 + b\lambda + c = 0$ has roots
$$\lambda = \frac{-b \pm \sqrt{(b^2 - 4ac)}}{2a} = -\frac{b}{2a} \text{ and } -\frac{b}{2a}.$$
Thus
$$y = A\exp\left(-\frac{bx}{2a}\right) + B\exp\left(-\frac{bx}{2a}\right)$$
$$= (A + B)\exp\left(-\frac{bx}{2a}\right).$$

This is undoubtedly a solution, but it has only *one* independent arbitrary constant, and it is therefore not the general solution. To obtain the general solution, we write

$$y = \exp\left(-\frac{bx}{2a}\right).u,$$

and examine what equation must be satisfied by u.† By differentiation, we see that

$$y = \exp\left(-\frac{bx}{2a}\right)u' - \frac{b}{2a}\exp\left(-\frac{bx}{2a}\right)u$$

and

$$y'' = \exp\left(-\frac{bx}{2a}\right)u'' - \frac{b}{a}\exp\left(-\frac{bx}{2a}\right)u' + \frac{b^2}{4a^2}\exp\left(-\frac{bx}{2a}\right)u.$$

After substitution into equation (5.2), and cancellation of the factor

$$\exp\left(-\frac{bx}{2a}\right),$$

we find that

$$a\left(u'' - \frac{b}{a}u' + \frac{b^2}{4a^2}u\right) + b\left(u' - \frac{b}{2a}u\right) + cu = 0,$$

or

$$au'' = 0.$$

Upon integrating twice, it follows that

$$u = Ax + B,$$

and so

$$y = (Ax + B)\exp\left(-\frac{bx}{2a}\right). \tag{5.7}$$

---

† We may anticipate that the equation for u will be second-order linear with constant coefficients, say $\alpha u' + \beta u' + \gamma u = 0$. However, it must certainly have a solution $u = $ const., so γ must be zero. Thus the equation for u is expected to be of the form $\alpha u'' + \beta u' = 0$, which is a first-order equation for u'. In the event it transpires that β is also zero and hence the equation is simply $u'' = 0$.

This solution, having two arbitrary constants, is the general solution.

Example 5.1.2

$$y'' + 2y' + y = 0.$$

Try $y \propto e^{\lambda x}$ → $\lambda^2 + 2\lambda + 1 = 0$.

That is,

$$(\lambda + 1)^2 = 0.$$

Therefore

$$\lambda = -1 \text{ or } -1.$$

The general solution is thus

$$y = (A + Bx)\,e^{-x}.$$

Case 3. Complex roots

We may write

$$\lambda = \frac{-b \pm \sqrt{(b^2 - 4ac)}}{2a} = -\frac{b}{2a} \pm i\beta \text{ say,}$$

where β is real. Thus

$$y = A \exp\left[\left\{-\frac{b}{2a} + i\beta\right\}x\right] + B \exp\left[\left\{-\frac{b}{2a} - i\beta\right\}x\right]$$

$$= \exp\left(-\frac{b}{2a}x\right)\{A\,e^{i\beta x} + B\,e^{-i\beta x}\}.$$

Now once we allow the use of complex roots we must logically allow the constants A and B to be complex also. On the other hand our final answer must be real, and accordingly we choose A and B to be complex conjugates. Thus if

$$A = A_1 + iA_2 \quad \text{and} \quad B = A_1 - iA_2,$$

we have

$$y = \exp\left(-\frac{b}{2a}x\right)\{A_1(e^{i\beta x} + e^{-i\beta x}) + iA_2(e^{i\beta x} - e^{-i\beta x})\}$$

$$= \exp\left(-\frac{b}{2a}x\right)\{2A_1 \cos \beta x - 2A_2 \sin \beta x\}$$

$$= \exp\left(-\frac{b}{2a}x\right)(C \cos \beta x + D \sin \beta x) \text{ say.} \qquad (5.8)$$

Accordingly we see that the real part of λ produces the genuinely exponential term, whereas the imaginary part produces the trigonometric terms. Again, we note that equation (5.8) is a solution of the equation and, having two arbitrary constants, is the general solution.

Example 5.1.3

$$y'' + y' + y = 0.$$

Try $y \propto e^{\lambda x} \rightarrow \lambda^2 + \lambda + 1 = 0$.

Thus

$$(\lambda + \tfrac{1}{2})^2 = -\tfrac{3}{4},$$

so

$$\lambda + \tfrac{1}{2} = \pm i\frac{\sqrt{3}}{2}.$$

Therefore

$$\lambda = -\tfrac{1}{2} \pm i\frac{\sqrt{3}}{2}.$$

Hence the general solution is

$$y = \exp(-\tfrac{1}{2}x)\{A \cos \tfrac{1}{2}\sqrt{(3)}x + B \sin \tfrac{1}{2}\sqrt{(3)}x\}$$

or, alternatively,

$$y = C \exp(-\tfrac{1}{2}x) \cos(\tfrac{1}{2}\sqrt{(3)}x + \varepsilon),$$

where C and ε are the arbitrary constants.

We have now seen how to obtain the complementary function. The next step is how to find a particular integral. There are a number of cases for which this can be done systematically, and we shall consider these in turn.

5.2   Calculating a particular integral: polynomial case

Suppose the function $f(x)$ on the right-hand side of equation (5.1) is a polynomial, so that

$$ay'' + by' + cy = A_0 + A_1x + A_2x^2 + \cdots + A_nx^n.$$

It is natural to examine whether there is a solution in which y is also a polynomial, so we try

$$y = B_0 + B_1 x + B_2 x^2 + \cdots + B_n x^n,$$
$$y' = \qquad B_1 + 2B_2 x + \cdots + nB_n x^{n-1},$$
$$y'' = \qquad\qquad 2B_2 + \cdots + n(n-1)B_n x^{n-2}.$$

Upon substituting into the equation and comparing coefficients of powers of x (beginning with x^n) it is found successively that

$$cB_n = A_n,$$
$$cB_{n-1} + bnB_n = A_{n-1},$$
$$cB_{n-2} + b(n-1)B_{n-1} + an(n-1)B_n = A_{n-2}, \text{ etc.}$$

Provided that $c \neq 0$ we can accordingly solve in turn for $B_n, B_{n-1}, \ldots, B_0$, and a particular integral has thus been found.

If $c = 0$, this means that the equation may be treated as a first-order equation for y'. Thus if we write $y' = Y$, we have

$$aY' + bY = \text{polynomial of degree } n.$$

Provided that $b = 0$ this has a particular integral in which Y is a polynomial of degree n, and so y is a polynomial of degree $n + 1$. In other words, we find a PI for the original equation for y by trying

$$y = B_0 + B_1 x + B_2 x^2 + \cdots + B_n x^n + B_{n+1} x^{n+1}.$$

Example 5.2.1

$$y'' + 2y' + y = x^2 + 4x + 7.$$

To find a PI we try

$$y = ax^2 + bx + c,$$
$$y' = 2ax + b,$$
$$y'' = 2a.$$

Hence

$$2a + 2(2ax + b) + (ax^2 + bx + c) \equiv x^2 + 4x + 7.$$

Therefore

$$a = 1,$$
$$4a + b = 4 \;\rightarrow\; b = 0,$$
$$2a + 2b + c = 7 \;\rightarrow\; c = 5.$$

It follows that $y = x^2 + 5$ is a PI. To obtain the general solution, we must add on the CF.

Example 5.2.2

$$y'' + 3y' = x^2 + 3x + 2.$$

We note that there is no term in y on the left-hand side of this equation. It follows that the relevant substitution is to try

$$y = ax^3 + bx^2 + cx + d,$$

which is one degree higher than the polynomial on the right-hand side of the equation. It follows that

$$y' = 3ax^2 + 2bx + c,$$

and that

$$y'' = 6ax + 2b,$$

so that substitution yields

$$6ax + 2b + 3(3ax^2 + 2bx + c) \equiv x^2 + 3x + 2.$$

Upon comparing coefficients of x^2, x^1, and x^0 we find successively that

$$9a = 1 \;\rightarrow\; a = \tfrac{1}{9},$$
$$6a + 6b = 3 \;\rightarrow\; b = \tfrac{7}{18},$$

and

$$2b + 3c = 2 \;\rightarrow\; c = \tfrac{11}{27}.$$

Thus

$$y = \tfrac{1}{9}x^3 + \tfrac{7}{18} x^2 + \tfrac{11}{28}x + d$$

is a particular integral for *any* value of x; we may take $d = 0$ for simplicity.

5.3   Calculating a particular integral: exponential case

Let us now consider the equation

$$ay'' + by' + cy = A\, e^{\lambda x}. \qquad (5.9)$$

It is natural to ask whether there is a solution of the form

$$y = B\, e^{\lambda x},$$

for a suitably chosen value of B. Upon substitution it is found that

$$a\lambda^2 B e^{\lambda x} + b\lambda B\, e^{\lambda x} + cB\, e^{\lambda x} \equiv A\, e^{\lambda x},$$

so that

$$B(a\lambda^2 + b\lambda + c) = A.$$

Provided $a\lambda^2 + b\lambda + c \neq 0$, this yields a value of B, and a particular integral has accordingly been found.

We note that the method breaks down whenever the value of λ on the right-hand side of equation (5.9) happens to be such that $a\lambda^2 + b\lambda + c = 0$, that is when the term on the right-hand side happens to be part of the complementary function. When this is the case, we write

$$y = e^{\lambda x}\, u,$$

and set up an equation for u. Substitution into equation (5.9) yields

$$a\, e^{\lambda x}(u'' + 2\lambda u' + \lambda^2 u) + b\, e^{\lambda x}(u' + \lambda u) + c\, e^{\lambda x}u = A\, e^{\lambda x},$$

or

$$au'' + (2a\lambda + b)u' + (a\lambda^2 + b\lambda + c)u = A,$$

that is,

$$au'' + (2a\lambda + b)u' = A.$$

This is a linear first-order equation for u' which can always be solved, as was shown in Chapter 2.

Example 5.3.1

$$y'' + 3y' + 2y = 2\, e^x.$$

To find a PI we try $y = A\, e^x$.

Therefore

$$A(1 + 3 + 2) = 2 \; \rightarrow \; A = \tfrac{1}{3}.$$

Thus

$$y = \tfrac{1}{3} \, e^x \text{ is a PI.}$$

Example 5.3.2

$$y'' + 3y' + 2y = e^{-x}.$$

To find a PI we try $y = A \, e^{-x}$.

Therefore

$$A(1 - 3 + 2) = 2.$$

We note that it is not possible to satisfy this equation. The reason for the difficulty is that e^{-x} is part of the complementary function, which can easily be shown to be $B \, e^{-x} + C \, e^{-2x}$. We accordingly write

$$y = e^{-x}u \; \rightarrow \; y' = e^{-x}(u' - u),$$
$$y'' = e^{-x}(u'' - 2u' + u).$$

Thus

$$(u'' - 2u' + u) + 3(u' - u) + 2u = 1,$$

or

$$u'' + u' = 1. \tag{5.10}$$

This linear equation for u' has an integrating factor e^x. Thus

$$\frac{\mathrm{d}}{\mathrm{d}x}(u' \, e^x) = e^x.$$

Therefore

$$u' \, e^x = e^x + a,$$
$$u' = 1 + a \, e^{-x},$$

and

$$u = x + b - a \, e^{-x}.$$

Finally

$$y = e^{-x}u = x \, e^{-x} + b \, e^{-x} - a \, e^{-2x}.$$

We note that this is the general solution for y, and not just a particular integral. The point here is that, having made the substitution $y = e^{-x}u$, we obtained equation (5.10), which we were able to integrate outright, without separate calculations of a particular integral and the complementary function.

5.4   Calculating a particular integral: oscillatory case

Let us now consider the procedure when equation (5.1) takes the form

$$ay'' + by' + cy = P \cos nx. \qquad (5.11)$$

We can always find a PI for this equation by trying

$$y = A \cos nx + B \sin nx,$$

substituting into equation (5.11), equating coefficients of $\cos nx$ and of $\sin nx$, and solving for A and B. This works well enough when a, b, c are simple numbers, but is not to be recommended in general.† A better method is to use complex variables. We replace y by y_1 (temporarily), so that

$$ay_1'' + by_1' + cy_1 = P \cos nx, \qquad (5.12)$$

and define a second function y_2 by the equation

$$ay_2'' + by_2' + cy_2 = P \sin nx. \qquad (5.13)$$

We multiply equations (5.12) and (5.13) by 1 and i, respectively, and add. This yields

$$az'' + bz' + c'z = P e^{inx},$$

where

$$z = y_1 + iy_2.$$

We have, by this device, converted our equation into one with an exponential term on the right-hand side. To find a PI we try

$$z = Z_0 e^{inx},$$

so that

$$z' = Z_0 \, in \, e^{inx} \quad \text{and} \quad z'' = Z_0(-n^2) \, e^{inx}.$$

† There is a tendency for students to ignore this remark—they are ill-advised to do so!

Thus we require that

$$Z_0\{(c - an^2) + ibn\}\, e^{inx} = P\, e^{inx},$$

that is,

$$Z_0 = \frac{P}{c - an^2 + ibn}.$$

Finally,

$$z = \frac{P\, e^{inx}}{c - an^2 + ibn}, \qquad (5.14)$$

and the solution we seek is y (or y_1) which is simply the real part of z.

To extract the real part in equation (5.14) we write

$$c - an^2 + ibn = \{(c - an^2)^2 + b^2n^2\}^{\frac{1}{2}}\, e^{i\varepsilon},$$

where

$$\tan \varepsilon = \frac{bn}{c - an^2}.$$

Thus

$$z = \frac{P}{\{(c - an^2)^2 + b^2n^2\}^{\frac{1}{2}}}\, e^{i(nx - \varepsilon)},$$

and hence

$$y = \frac{P}{\{(c - an^2)^2 + b^2n^2\}^{\frac{1}{2}}}\, \cos (nx - \varepsilon). \qquad (5.15)$$

We note that, overall, the procedure is as follows. (a) Replace $\cos nx$ by a complex exponential of which it is the real part. (b) Solve the amended equation. (c) Take the real part of the answer.

Example 5.4.1

$$y'' + y' + 2y = \cos x + \sin x$$

To find a PI, we may add together PI's corresponding to right-hand sides of $\cos x$, $\sin x$, respectively. These can both be obtained by consideration of the equation

$$y'' + y' + 2y = e^{ix}.$$

Try $y = A\,e^{ix}$, so that $y' = i\,A\,e^{ix}$ and $y'' = -A\,e^{ix}$.
Therefore

$$A(-1 + i + 2)\,e^{ix} = e^{ix},$$

therefore

$$A = \frac{1}{1 + i} = \frac{1}{\sqrt{2}}\exp\left(-i\frac{\pi}{4}\right),$$

and

$$y = \frac{1}{\sqrt{2}}\exp\left\{i\left(x - \frac{\pi}{4}\right)\right\}.$$

Thus, taking real parts, the PI for the equation

$$y'' + y' + 2y = \cos x$$

is

$$y = \frac{1}{\sqrt{2}}\cos\left(x - \frac{\pi}{4}\right) = \tfrac{1}{2}(\cos x + \sin x).$$

Likewise, taking imaginary parts, the PI for the equation

$$y'' + y' + 2y = \sin x$$

is

$$y = \frac{1}{\sqrt{2}}\sin\left(x - \frac{\pi}{4}\right) = \tfrac{1}{2}(\sin x - \cos x).$$

Adding together these two PI's, we see that a PI for the original equation is

$$y = \sin x.$$

5.5   Concluding general remarks

Two general points may now be usefully made. First of all it is perhaps as well to remind the reader that it is not possible to determine the values of the two arbitrary constants until the complete solution has been obtained. Many students make the mistake of obtaining the CF, which involves two arbitrary constants, and then applying the boundary conditions to the complementary function alone. It is essential to add on a particular integral first, and only then should the boundary conditions be imposed.

Example 5.5.1

Solve $3y'' + y' + 2y = \cos x$, subject to the conditions $y = y' = 0$ when $x = 0$.

To obtain the CF, we try $y = A e^{\lambda x}$, whence
$$3\lambda^2 + \lambda + 2 = 0.$$

Therefore
$$\lambda^2 + \tfrac{1}{3}\lambda = -\tfrac{2}{3},$$

so
$$(\lambda + \tfrac{1}{6})^2 = -\tfrac{2}{3} + \tfrac{1}{36} = -\tfrac{23}{36},$$

and therefore
$$\lambda = -\tfrac{1}{6} \pm \tfrac{1}{6}\sqrt{(23)}\mathrm{i}.$$

Hence the CF is
$$y_{\mathrm{CF}} = \exp\left(-\tfrac{1}{6}x\right)\{A \cos \tfrac{1}{6}\sqrt{(23)}x + B \sin \tfrac{1}{6}\sqrt{(23)}x\}.$$

To obtain a PI we write
$$3y'' + y' + 2y = e^{\mathrm{i}x},$$

and try
$$y = C e^{\mathrm{i}x}.$$

Thus
$$(\mathrm{i} - 1)C = 1.$$

Therefore
$$C = \frac{1}{\mathrm{i} - 1} = \frac{1}{\sqrt{2} \exp\left(\tfrac{3}{4}\pi\mathrm{i}\right)},$$

and
$$y_{\mathrm{PI}} = \frac{1}{\sqrt{2}} \exp\{\mathrm{i}(x - \tfrac{3}{4}\pi)\}$$

$$= \frac{1}{\sqrt{2}} \cos(x - \tfrac{3}{4}\pi) \quad \text{(on taking the real part)}$$

$$= \tfrac{1}{2}(\sin x - \cos x).$$

Adding, we have

$$y = y_{CF} + y_{PI} = \exp(-\tfrac{1}{6}x)\{A\cos\tfrac{1}{6}\sqrt{(23)}x + B\sin\tfrac{1}{6}\sqrt{(23)}x\}$$
$$+ \tfrac{1}{2}(\sin x - \cos x).$$

When $x = 0$,

$$y = 0 = A - \tfrac{1}{2}.$$

Therefore

$$A = \tfrac{1}{2}.$$

Differentiating, we have

$$y' = \exp(-\tfrac{1}{6}x)\{-A\sin\tfrac{1}{6}\sqrt{(23)}x + B\cos\tfrac{1}{6}\sqrt{(23)}x\}$$
$$\times \tfrac{1}{6}\sqrt{(23)} - \tfrac{1}{6}\exp(-\tfrac{1}{6}x)$$
$$\times \{A\cos\tfrac{1}{6}\sqrt{(23)}x + B\sin\tfrac{1}{6}\sqrt{(23)}x\}$$
$$+ \tfrac{1}{2}(\cos x + \sin x).$$

Thus, when $x = 0$

$$y' = 0 = \tfrac{1}{6}\sqrt{(23)}B - \tfrac{1}{6}A + \tfrac{1}{2},$$

and so

$$\sqrt{(23)}B = A - 3 = -\tfrac{5}{2}.$$

Hence, finally,

$$y = \tfrac{1}{2}(\sin x - \cos x)$$
$$+ \exp(-\tfrac{1}{6}x)\left\{\tfrac{1}{2}\cos\tfrac{1}{6}\sqrt{(23)}x - \frac{5}{2\sqrt{(23)}}\sin\tfrac{1}{6}\sqrt{(23)}x\right\}.$$

Having seen how to solve the linear second-order equation (5.1) when $f(x)$ is polynomial, exponential, or sinusoidal, we may reasonably ask what may be done in other cases. The observant student may already have some ideas as to the answer to this question. Take the equation (5.1) as it stands,

$$ay'' + by' + cy = f(x).$$

We know that the CF is

$$y = A\exp(\lambda_1 x) + B\exp(\lambda_2 x),$$

where λ_1 and λ_2 are the roots of $a\lambda^2 + b\lambda + c = 0$. We write

$$y = e^{\lambda x}u,$$

where λ is either of the roots λ_1, λ_2, and substitute. This yields

$$e^{\lambda x}\{au'' + (2a\lambda + b)u' + (a\lambda^2 + b\lambda + c)u\} = f(x),$$

or

$$u'' + \left(2\lambda + \frac{b}{a}\right)u' = \frac{1}{a}f(x)\,\mathrm{e}^{-\lambda x}.$$

This is now a first-order linear equation for u', having an integrating factor

$$\exp\left\{\left(2\lambda + \frac{b}{a}\right)x\right\},$$

so that

$$\frac{\mathrm{d}}{\mathrm{d}x}\left[u'\exp\left\{\left(2\lambda + \frac{b}{a}\right)x\right\}\right] = \frac{1}{a}f(x)\exp\left\{\left(\lambda + \frac{b}{a}\right)x\right\},$$

and thus

$$u'\exp\left\{\left(2\lambda + \frac{b}{a}\right)x\right\}$$
$$= A + \frac{1}{a}\int^{x} f(x)\exp\left\{\left(\lambda - \frac{b}{a}\right)x\right\}\mathrm{d}x. \qquad (5.16)$$

The integral in equation (5.16) can always be evaluated numerically, even when an explicit evaluation is not possible, so in either case u' is known, and u follows by a further integration.

Tutorial Examples

1. Solve the following differential equations, giving the general solution in each case.

    (a) $y'' + 7y' + 12y = 0.$
    (b) $y'' + 4y' + 4y = 0.$
    (c) $4y'' + 5y' + 2y = 0.$
    (d) $y'' + y' - 2y = 0.$

2. Obtain particular integrals for each of the following equations.

    (a) $y'' + 2y' + 5y = 1 + x^2.$
    (b) $2y'' + 3y' = x + \tfrac{1}{2}x^2.$

    (c) $y'' + 3y' + 4y = \sin 2x$.

    (d) $y'' + 4y = \cos 2x$.

    (e) $y'' + 3y' + 2y = e^{-x} + 2e^{-2x}$.

3. Solve the following equations, subject to the given boundary
conditions.

    (a) $2y'' + 5y' + 4y = x^2 + x + 1$,
        $y = 1$, $y' = 0$ when $x = 0$.

    (b) $y'' - 4y = \cos 2x + \sin 2x$,
        $y = 0$ when $x = 0$,
        $y = 1$ when $x = \pi/2$.

    (c) $y'' + 4y' + 4y = 2\,e^x$,
        $y = 2$ when $x = 0$,
        $y = 0$ when $x = 2$.

4. Obtain the general solution of the following equations.

    (a) $y'' - y = \dfrac{2}{1 + e^x}$.

    (b) $3y'' + 7y' + 4y = x^2\,e^x$.

CHAPTER 6

Applications of Second-Order Equations

6.1 Free oscillations of a mechanical system

Consider a mass m attached to a fixed support by means of a light vertical spring. The tension in the spring is taken to be directly proportional to its extension, the constant of proportionality k being called the spring constant. Suppose the natural length of the spring is l_0 (see Fig. 6.1). When the mass

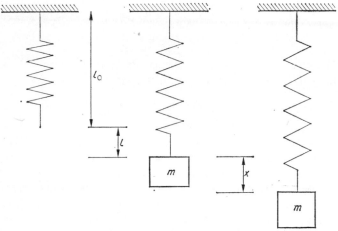

FIG. 6.1 Mass–spring system

is attached, the spring will be extended by an additional amount l when in equilibrium, where

$$kl = mg, \qquad (6.1)$$

since the upward tension in the spring must balance the downward force due to gravity acting on the mass.

We now imagine the mass to be displaced vertically from its equilibrium position. We measure x downwards from the position of equilibrium. Then Newton's law shows that

$$\frac{d}{dt}\left(m\,\frac{dx}{dt}\right) = \text{Rate of change of downward momentum}$$

$$= \text{Net downward force}$$

$$= mg - T$$

$$= mg - k(l + x),$$

since the extension of the spring is now $l + x$, and upon making use of equation (6.1) this yields

$$m\,\frac{d^2x}{dt^2} = -kx,$$

or

$$\frac{d^2x}{dt^2} + \omega_0^2 x = 0, \tag{6.2}$$

where

$$\omega_0^2 = \frac{k}{m}.$$

The general solution of equation (6.2) is

$$x = A\cos\left(\omega_0 t + \varepsilon\right).$$

We note that the solution represents an oscillation of period $2\pi/\omega_0$—that is, when t increases by $2\pi/\omega_0$, $\omega_0 t + \varepsilon$ increases by 2π, and so the value of x repeats itself. The quantity ω_0 is called the *natural* frequency of the system, in the sense that vibrations of this frequency can be maintained (in the absence of damping) independently of any externally imposed forcing agency. The quantity A, which tells us how far the mass vibrates, is called the *amplitude* of the oscillation.

As an illustrative example, consider a cage of mass m which is being pulled upwards with uniform speed u by a long steel cable. At time $t = 0$, the upper end of the cable is suddenly fixed. If k is the stiffness of the spring, that is, of the cable, and if x is measured upwards from the position of the mass at time $t = 0$, then when the mass has risen a distance x there will be a compression in the cable such as to produce a net *downward* force kx on the mass. Thus

$$m\ddot{x} + kx = 0.$$

Therefore

$$x = A \cos (\omega_0 t + \varepsilon). \tag{6.3}$$

The initial conditions at $t = 0$ are that $x = 0$ and $\dot{x} = u$ Upon substitution in equation (6.3) these yield

$$A \cos \varepsilon = 0$$

and

$$-A\omega_0 \sin c = u,$$

so that

$$\varepsilon = \frac{\pi}{2}$$

and

$$A = - \frac{u}{\omega_0}.$$

The solution (6.3) accordingly becomes

$$x = \frac{u}{\omega_0} \sin \omega_0 t.$$

We see that the amplitude of the oscillation of the cage is

$$\frac{u}{\omega_0} = u \sqrt{\frac{k}{m}}.$$

6.2 Free oscillations with damping

Suppose we now add a damping mechanism to the mass-spring system. This, in its simplest form, consists of a plunger in a tube of viscous liquid (see Fig. 6.2), which resists any relative motion by exerting a resistive force proportional to the

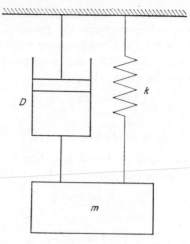

FIG. 6.2   Mass–spring–damper system

relative velocity, and is called a dashpot. The constant of proportionality will simply be called the dashpot constant.

As before we measure x downwards from the equilibrium position, so that the downward force mg due to gravity is balanced by the tension already in the spring in equilibrium. The equation of motion is then

$$\frac{d}{dt}\left(m\,\frac{dx}{dt}\right) = \text{Net downward force}$$

$$= -kx - D\dot{x},$$

or

$$m\ddot{x} + D\dot{x} + kx = 0.$$

This second-order linear equation with constant coefficients has solutions of the form

$$x \propto e^{\lambda t},$$

with

$$m\lambda^2 + D\lambda + k = 0,$$

that is,

$$\lambda = \frac{-D \pm \sqrt{(D^2 - 4km)}}{2m}.$$

The nature of the solution clearly depends upon the sign of $D^2 - 4km$.

When $D^2 > 4km$, we have what is called *heavy* damping. The two values of λ are both real, and are both negative, since D is positive and greater than $\sqrt{(D^2 - 4km)}$. Thus

$$x = A \exp(\lambda_1 t) + B \exp(\lambda_2 t), \quad (\lambda_1 < 0, \lambda_2 < 0),$$

and any disturbance will decay exponentially to zero.

When $D^2 < 4km$, we have *light* damping. The two values of λ are then both complex, and we may write

$$\lambda = \frac{-D \pm i\sqrt{(4km - D^2)}}{2m}$$

$$= -\frac{D}{2m} \pm i\omega,$$

where

$$\omega^2 = \frac{4km - D^2}{4m^2} = \frac{k}{m} - \frac{D^2}{4m^2}$$

$$= \omega_0^2 - \frac{D^2}{4m^2}. \tag{6.4}$$

The solution then becomes

$$x = A \exp\left(-\frac{Dt}{2m}\right) \cos(\omega t + \varepsilon). \tag{6.5}$$

We note that this solution is oscillatory, but that the amplitude of the oscillation decays exponentially with time. The frequency, ω, of the oscillation is given by equation (6.4), from which it is seen that the effect of damping is to *reduce* the frequency. However, if D is small D^2 will be *very* small, and the frequency ω will be close to the natural frequency ω_0.

It is fairly easy to show that the successive maxima of the oscillation form a geometric progression. For example, we deduce from equation (6.5) that

$$\frac{dx}{dt} = -A \exp\left(-\frac{Dt}{2m}\right)\left\{\frac{D}{2m} \cos(\omega t + \varepsilon) + \omega \sin(\omega t + \varepsilon)\right\}.$$

This is zero, and hence x has a maximum or minimum, whenever

$$\tan(\omega t + \varepsilon) = -\frac{D}{2m\omega},$$

and so

$$\omega t + \varepsilon = -\tan^{-1}\left(\frac{D}{2m\omega}\right)$$

$$= \omega t_0 + n\pi \text{ say.}$$

Thus the maxima and/or minima of x are given by

$$x = A \exp\left\{-\frac{D}{2m\omega}(\omega t_0 - \varepsilon + n\pi)\right\} \cos(\omega t_0 + n\pi)$$

$$= A_1 \exp\left(-\frac{D\pi}{2m\omega}n\right)(-1)^n \cos \omega t_0.$$

Clearly maxima and minima alternate (as common sense tells us!), and the magnitudes of successive turning values are in the ratio

$$\frac{\exp\left(-\frac{D\pi}{2m\omega}(n+1)\right)}{\exp\left(-\frac{D\pi}{2m\omega}n\right)} = \exp\left(-\frac{D\pi}{2m\omega}\right).$$

Finally, when $D^2 = 4km$, we have *critical* damping. The two values of λ are each equal to $-D/2m$, and the solution is

$$x = (A + Bt) \exp\left(-\frac{Dt}{2m}\right).$$

This solution is not oscillatory. However, when $A = 0$ and $B = 1$ (for example) the value of x increases with t for small t, and then decays to zero at large t. A graph of x against t accordingly gives the impression of an attempted oscillation which fizzles out like a damp squib!

6.3 Forced oscillations without damping

Consider now a mass m, which is attached by a spring to a movable support. Suppose the support is at rest when $t < 0$, and that the mass is in equilibrium. When $t > 0$ the support moves so that the downward displacement is $\xi(t)$. As before, we measure x downwards from the equilibrium position. Then the mass moves downwards a distance x, and the support moves downwards a distance ξ, so the spring is further stretched an amount $x - \xi$. The equation of motion is accordingly

$$\frac{\mathrm{d}}{\mathrm{d}t}\left(m\frac{\mathrm{d}x}{\mathrm{d}t}\right) = \text{net downward force}$$

$$= -k(x - \xi),$$

so

$$m\frac{\mathrm{d}^2x}{\mathrm{d}t^2} + kx = k\xi,$$

or

$$\frac{\mathrm{d}^2x}{\mathrm{d}t^2} + \omega_0^2 x = \omega_0^2 \xi. \tag{6.6}$$

The solution of equation (6.6) consists of the sum of the CF and a PI. The CF is simply

$$x_{\mathrm{CF}} = A\cos(\omega_0 t + \varepsilon). \tag{6.7}$$

The PI, of course, depends upon the precise way in which the support moves, that is, upon ξ. If the support vibrates so that

$$\xi = h\sin nt,$$

then we write

$$\frac{\mathrm{d}^2x}{\mathrm{d}t^2} + \omega_0^2 x = -\,\mathrm{i}\omega_0^2 h\,\mathrm{e}^{\mathrm{i}nt}, \tag{6.8}$$

and look for a solution

$$x_{\mathrm{PI}} = X\,\mathrm{e}^{\mathrm{i}nt}.$$

Thus

$$-n^2 X\,\mathrm{e}^{\mathrm{i}nt} + \omega_0^2 X\,\mathrm{e}^{\mathrm{i}nt} = -\mathrm{i}h\omega_0^2\,\mathrm{e}^{\mathrm{i}nt},$$

or

$$(\omega_0^2 - n^2)X = -ih\omega_0^2.$$

Provided that $\omega_0^2 - n^2 \neq 0$, that is, provided that the forcing frequency n is not equal to the natural frequency ω_0, we have

$$X = \frac{-ih\omega_0^2}{\omega_0^2 - n^2},$$

so

$$x_{PI} = -\frac{ih\omega_0^2}{\omega^2 - n^2} e^{int}$$

$$= \frac{h\omega_0^2}{\omega_0^2 - n^2} \sin nt, \qquad (6.9)$$

after taking the real part.

Upon adding together the results (6.7) and (6.9) we have

$$x = A \cos(\omega_0 t + \varepsilon) + \frac{h\omega_0^2}{\omega_0^2 - n^2} \sin nt.$$

The arbitrary constants, A and ε, are determined by the initial conditions, which could well be that $x = (dx/dt) = 0$ when $t = 0$.

We note that the part of the solution corresponding to the CF, given by equation (6.7), has the same frequency ω_0 as the natural frequency of the system. This part of the solution is referred to as a free vibration. The remainder, given by equation (6.9), has the same frequency n as the forcing term, and is referred to as a forced vibration. When damping is present, as it must always be in reality, the free vibrations will die out with time, and only the forced vibrations will persist.

It remains to examine the form of the forced vibrations when $n = \omega_0$. In this case there is not a PI of equation (6.8) having the form $x_{CF} = X e^{int}$. We accordingly write

$$\frac{d^2x}{dt^2} + n^2x = -in^2h\,e^{int},$$

and substitute

$$x = e^{int}\,u.$$

This leads to

$$\ddot{u} + 2in\dot{u} = -in^2h,$$

so that

$$\frac{\mathrm{d}}{\mathrm{d}t}(\dot{u}\,e^{2int}) = -in^2h\,e^{2int}.$$

Therefore

$$\dot{u}\,e^{2int} = A - \tfrac{1}{2}nh\,e^{2int},$$
$$\dot{u} = A\,e^{-2int} - \tfrac{1}{2}nh,$$

and

$$u = A_1\,e^{-2int} - \tfrac{1}{2}nht + B.$$

Thus

$$x = A_1\,e^{-int} + B\,e^{int} - \tfrac{1}{2}nht\,e^{int}$$
$$= C\cos(nt + \varepsilon) - \tfrac{1}{2}nht\cos nt.$$

We note that the forced oscillations grow with time. Also, since

$$-\cos nt = \sin(nt - \pi/2),$$

we see that they lag behind the forcing term in phase by an amount $\pi/2$.

6.4 Forced oscillations with damping

Consider now a mass m, suspended from a fixed support by a spring of stiffness k and a dashpot of damping coefficient D. The mass is subjected directly to a downward forcing term $P = P_0 \cos nt$. Then

$$\frac{\mathrm{d}}{\mathrm{d}t}\left(m\,\frac{\mathrm{d}x}{\mathrm{d}t}\right) = \text{Net downward force}$$

$$= P - kx - D\dot{x},$$

or

$$m\ddot{x} + D\dot{x} + kx = P = P_0 \cos nt. \qquad (6.10)$$

The free vibrations, represented by the CF, are as discussed in Section 6.2. These, as we have already noted, die out with time, whereas the forced oscillations persist. To find the forced oscillations we write

$$m\ddot{x} + D\dot{x} + kx = P_0\, e^{int},$$

and try

$$x = X\, e^{int}.$$

It then follows that

$$X(k + Dni - mn^2) = P_0,$$

and so

$$x = \frac{P_0}{k - mn^2 + Dni}\, e^{int}. \qquad (6.11)$$

To take real parts we write

$$k - mn^2 + Dn\, i = |\, k - mn^2 + Dn\, i\,|\, e^{i\varepsilon},$$

where

$$\tan \varepsilon = \frac{Dn}{k - mn^2}. \qquad (6.12)$$

Then equation (6.11) becomes

$$x = \frac{P_0}{\{(k - mn^2)^2 + D^2n^2\}^{\frac{1}{2}}}\cos{(nt - \varepsilon)}. \qquad (6.13)$$

The amplitude of the forced oscillations is thus

$$A = |\, X\,| = \frac{P_0}{\{(k - mn^2)^2 + D^2n^2\}^{\frac{1}{2}}}. \qquad (6.14)$$

It is instructive to express this result in an essentially non-dimensional form, and accordingly we write

$$k = m\omega_0^2$$

and

$$D = b\,.\,D_c = 2b\sqrt{(mk)}.$$

Then equation (6.14) becomes

$$A = \frac{P_0}{\left\{\left(1 - \dfrac{n^2}{\omega_0^2}\right)^2 + 4b^2\,\dfrac{n^2}{\omega_0^2}\right\}^{\frac{1}{2}}}$$

$$= \frac{P_0}{k}\{(1 - z)^2 + 4b^2z\}^{-\frac{1}{2}}. \qquad (6.15)$$

It may be noted that, for a given system, the amplitude of the oscillation is a maximum when

$$(1 - z)^2 + 4b^2z$$

is a minimum. This occurs when

$$-2(1 - z) + 4b^2 = 0,$$

that is,

$$z = 1 - 2b^2,$$

that is,

$$n = \omega_0(1 - 2b^2)^{\frac{1}{2}}. \tag{6.16}$$

If b is small, that is, if the damping is small, the maximum amplitude occurs when the frequency of the forcing term is close to the natural frequency. This maximum amplitude is given by

$$\left(\frac{A_m k}{P_0}\right)^2 = \frac{1}{(1 - z)^2 + 4b^2z}$$

$$= \frac{1}{4b^4 + 4b^2(1 - 2b^2)}$$

$$= \frac{1}{4b^2(1 - b^2)},$$

so that

$$A_m = \frac{P_0}{2kb(1 - b^2)^{\frac{1}{2}}}. \tag{6.17}$$

When b is small, the maximum amplitude is large but is infinite only when the damping is ignored completely.

Finally, it may be deduced from equation (6.12) that when the system is vibrated at precisely its natural frequency, so that $k - mn^2 = 0$, the value of ε is $\pi/2$. Thus the oscillation lags behind the forcing term in phase by exactly $\pi/2$.

6.5   Electrical oscillations

Our study of practical applications of second-order ordinary differential equations has so far been confined to the case of mechanical oscillations, for which we have presented considerable detail in the preceding sections. The principles, however, apply equally well to oscillation of electrical circuits. Consider, for example, a simple electrical circuit consisting of a capacitance, an inductance, and a resistance, together with an applied e.m.f., as shown in Fig. 6.3. We have already seen

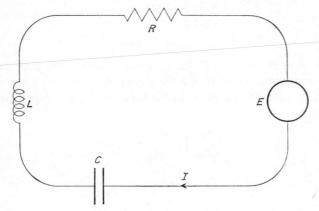

FIG. 6.3   A simple electrical circuit

that the potential drop across an inductance L is $L(\mathrm{d}I/\mathrm{d}t)$, and that across a resistance R is RI. The potential drop across a capacitance C is $(1/C)Q$, where Q is the charge on the capacitance. Thus the condition that the total potential drop around the circuit must be balanced by the applied e.m.f. gives

$$L\frac{\mathrm{d}I}{\mathrm{d}t} + RI + \frac{1}{C}Q = E(t). \qquad (6.18)$$

But

$$I = \frac{\mathrm{d}Q}{\mathrm{d}t},$$

and hence equation (6.18) may be written as

$$L\frac{\mathrm{d}^2Q}{\mathrm{d}t^2} + R\frac{\mathrm{d}Q}{\mathrm{d}t} + \frac{1}{C}Q = E(t), \qquad (6.19)$$

which is of precisely the same form as the equation (6.10) for forced mechanical oscillations with damping.

As an example we consider the case of an oscillatory applied e.m.f., for which

$$E(t) = E_0 \cos nt.$$

Following the ideas developed earlier, we write

$$\frac{1}{C} = L\omega_0^2 \quad \text{and} \quad R = 2b(L/C)^{\frac{1}{2}}, \tag{6.20}$$

so that equation (6.19) takes the somewhat simpler form

$$\ddot{Q} + 2b\omega_0\dot{Q} + \omega_0^2 Q = \frac{E_0}{L} e^{int}. \tag{6.21}$$

The CF, representing the free oscillations, contains only terms which die out as t increases. The forced oscillations are obtained by trying

$$Q = Q_0 e^{int},$$

whence

$$Q_0 = \frac{E_0/L}{\omega_0^2 - n^2 + 2b\omega_0 ni},$$

and

$$Q = \frac{E_0/L}{\omega_0^2 - n^2 + 2b\omega_0 n \, i} e^{int}$$

$$= \frac{E_0/L}{\{(\omega_0^2 - n^2)^2 + 4b^2\omega_0^2 n^2\}^{\frac{1}{2}}} \cos (nt - \varepsilon), \tag{6.22}$$

with

$$\tan \varepsilon = \frac{2b\omega_0 n}{\omega_0^2 - n^2}. \tag{6.23}$$

We may re-write equation (6.22) as

$$Q = CE_0 \left\{ \left(1 - \frac{n^2}{\omega_0^2}\right)^2 + 4b^2 \frac{n^2}{\omega_0^2} \right\}^{-\frac{1}{2}} \cos (nt - \varepsilon), \tag{6.24}$$

so the amplitude of the charge oscillation is

$$CE_0 \left\{ \left(1 - \frac{n^2}{\omega_0^2}\right)^2 + 4b^2 \frac{n^2}{\omega_0^2} \right\}^{-\frac{1}{2}}.$$

This may be shown, as before, to be a maximum when

$$n = \omega_0(1 - 2b^2)^{\frac{1}{2}}. \tag{6.25}$$

Upon differentiating equation (6.24) we deduce that

$$I = -CE_0 n \left\{ \left(1 - \frac{n^2}{\omega_0^2}\right)^2 + 4b^2 \frac{n^2}{\omega_0^2} \right\}^{-\frac{1}{2}} \sin(nt - \varepsilon), \tag{6.26}$$

so the amplitude of the current oscillation is

$$CE_0 n \left\{ \left(1 - \frac{n^2}{\omega_0^2}\right)^2 + 4b^2 \frac{n^2}{\omega_0^2} \right\}^{-\frac{1}{2}}. \tag{6.27}$$

This quantity has a maximum at a value of n different from that given by equation (6.25). If we write $n^2/\omega_0^2 = z$, it follows that the maximum occurs when

$$z\{(1 - z)^2 + 4b^2 z\}^{-1}$$

is a maximum, and hence when

$$z^{-1} + z + 4b^2 - 2$$

is a minimum, which shows that

$$z = 1.$$

In other words, the maximum amplitude of current oscillation occurs when the forcing frequency is precisely the natural frequency. We may deduce from equation (6.27) that this maximum amplitude is equal to

$$\frac{CE_0 \omega_0}{2b},$$

and it follows in turn from equation (6.20) that this is equal to

$$E_0/R.$$

Under these conditions equation (6.26) becomes

$$I = -\frac{E_0}{R} \sin(\omega_0 t - \varepsilon),$$

and since $\varepsilon = \pi/2$ when $n = \omega_0$, as equation (6.23) shows, this finally becomes

$$I = \frac{E_0}{R} \cos \omega_0 t.$$

In other words, when the forcing frequency is equal to the natural frequency the effects of the inductance and the capacitance exactly balance out.

Tutorial Examples

1. A particle of mass m, initially at rest at $x = a$, is attracted towards the origin by a force of magnitude mk^2x. When it reaches the point $x = \frac{1}{2}a$, the attractive force is replaced by a repelling one, again of magnitude mk^2x. Find how long it takes for the particle to reach the origin from $x = a$.

2. The springs of a car are compressed a m under its weight. It travels over a road which has a cosine wave profile of amplitude h m and wavelength λ m. Show that, if the damping is ignored, the amplitude of the forced vibrations will be

$$h\left\{1 - \left(\frac{2\pi v_0}{\lambda}\right)^2 \frac{a}{g}\right\}^{-1} \text{ m,}$$

when the car travels at v_0 m s^{-1}. Deduce that the worst vibrations occur when the speed of the car is

$$v_0 = \frac{\lambda}{2\pi}\left(\frac{g}{a}\right)^{\frac{1}{2}} \text{ m s}^{-1}$$

3. A car of mass m travels along a flat road at a steady speed v_0. The propulsive force exerted by the engine is P, and the overall resistance (air, road, etc.) may be assumed to be proportional to the square of the speed. At a certain time the thrust is removed, and the brakes are applied. If the effect of the brakes is roughly equivalent to applying a constant resistive force R, show that the subsequent motion of the car is governed by the equation

$$m\ddot{x} + \frac{P}{v_0^2}\dot{x}^2 = -R,$$

and that the distance taken to bring the car to rest is

$$x = \frac{mv_0^2}{2P} \log \left(1 + \frac{P}{R}\right).$$

4. A mass m is hung from a horizontal platform by a spring of stiffness k and a dashpot of damping constant D. The platform vibrates vertically, with downward displacement $h \cos nt$. Let x be measured downwards from the platform. Show that the amplitude A of the forced oscillations in x is given by

$$\frac{A}{h} = \frac{n^2}{\omega_0^2} \left\{ \left(1 - \frac{n^2}{\omega_0^2}\right)^2 + 4b^2 \frac{n^2}{\omega_0^2} \right\}^{-\frac{1}{2}}.$$

5. A mass m, subject to a vertical alternating force $P_0 \cos nt$, is mounted on a rigid horizontal platform by springs of stiffness k, and a dashpot with damping constant D is also inserted between the mass and the platform. Both the spring force and the damping force are transmitted to the foundation. Find the amplitude A of the transmitted force, and determine the condition that $A/P_0 < 1$.

6. Two identical cars are at rest at traffic lights. When the lights change car A sets off under maximum throttle, producing a propulsive force P, the overall resistance is proportional to the speed, and the car gradually accelerates towards its maximum speed v_0. The driver of car B adjusts the accelerator so that the force exerted by his engine is proportional to the speed of car A, reaching the maximum value P as car A reaches its maximum speed v_0. Show that the positions x, y of cars A and B satisfy the equations

$$m\ddot{x} + \frac{P}{v_0}\dot{x} = P,$$

$$m\ddot{y} + \frac{P}{v_0}\dot{y} = \frac{P}{v_0}\dot{x}.$$

Solve these equations and show that, after a sufficient time, the distance between the cars will be

$$mv_0^2/P.$$

7. The same two cars as in the previous problem stop at another set of traffic lights. This time car A sets off with uniform acceleration a. The driver of car B adjusts the accelerator so that the force exerted by his engine is proportional to the distance between the cars, maximum throttle being reached when the separation is l. Show that the equation of motion of car B is

$$m\ddot{y} + \frac{P}{v_0}\dot{y} + \frac{P}{l}y = \frac{Pa}{2l}t^2.$$

Deduce that the motion of car B will be non-oscillatory only if

$$l \geqslant \frac{4mv_0^2}{P}.$$

If l is chosen so that the equality holds in the above equation, show that

$$\ddot{y} + 2\alpha\dot{y} + \alpha^2 y = \tfrac{1}{2}a\alpha^2 t^2,$$

where

$$\alpha = \frac{P}{2mv_0},$$

and hence that

$$y = \tfrac{1}{2}at^2 - \frac{2a}{\alpha}t + \frac{3a}{\alpha^2} - \frac{a}{\alpha^2}e^{-\alpha t}(3 + \alpha t).$$

8. A simple electrical circuit consists of an inductance L, a resistance R, and a capacitance C in series. An e.m.f. of voltage $E_0 \cos \omega t$ is impressed on the circuit, having a frequency ω equal to the natural frequency of the circuit. Show that, after a sufficient time, the amplitudes of the oscillating potential differences across the inductance and the capacitance are each equal to

$$\frac{E_0}{R}\left(\frac{L}{C}\right)^{\frac{1}{2}}.$$

9. A simple electrical circuit consists of an inductance L, a resistance R, and a capacitance C in series, subject to an applied e.m.f. equal to

$$E_0\{\cos \omega_0 t + \varepsilon \cos 2\omega_0 t\},$$

where ω_0 is the natural frequency of the circuit. Calculate the current which flows after a large time, when the free oscillations have been damped out.

CHAPTER 7

Other Second-Order Equations

7.1 Equations from which x is missing

We have already seen that the most general second-order equation may be expressed functionally as

$$F(x, y, y', y'') = 0.$$

Accordingly, when x does not explicitly arise in the equation, we have

$$F(y, y', y'') = 0. \tag{7.1}$$

It is natural to enquire whether any useful purpose would be achieved by taking y as the new independent variable and $y' = p$ as the new dependent variable. In fact, as we shall now show, the effect of making this change of variables is to change the second-order equation into a first-order equation.

What we need to do is to replace x derivatives by y derivatives. We therefore write

$$\frac{d}{dx} \rightarrow \frac{d}{dy} \cdot \frac{dy}{dx} = p\frac{d}{dy},$$

and hence

$$y'' = \frac{d^2y}{dx^2} \rightarrow p\frac{dp}{dy}.$$

The equation (7.1) thus becomes

$$F\left(y, p, p\frac{dp}{dy}\right) = 0,$$

or

$$G\left(y, p, \frac{dp}{dy}\right) = 0,$$

which is a first-order equation between y and p. If this can be integrated by the methods of Chapter 2, our problem is solved.

Example 7.1.1

$$y(y-1)y'' + (y')^2 = 0.$$

We write $y' = p$ and $y'' = p(\mathrm{d}p/\mathrm{d}y)$, so the equation becomes

$$y(y-1)p\frac{\mathrm{d}p}{\mathrm{d}y} + p^2 = 0.$$

Either $p = 0$, so that

$$y = \text{const.,} \qquad (7.2)$$

or, upon cancelling by p,

$$y(y-1)\frac{\mathrm{d}p}{\mathrm{d}y} + p = 0,$$

which separates to give

$$\frac{\mathrm{d}p}{p} = -\frac{\mathrm{d}y}{y(y-1)}$$

$$= \frac{\mathrm{d}y}{y} - \frac{\mathrm{d}y}{y-1}.$$

Integration yields

$$\log|p| = \log|y| - \log|y-1| + \text{const.,}$$

so that

$$|p| = \frac{|y|}{|y-1|} \cdot \text{const.,}$$

and hence

$$p = \frac{ky}{y-1}.^{\dagger}$$

---

† By neglecting the modulus signs in this equation, we are effectively allowing the sign of k to change when y goes through the values 0 and 1. For example, if the boundary conditions are $y = 2$ and $y' = p = 1$ when $x = 0$, then $c = \log 2 - 2$ and $k = \frac{1}{2}$. This value of k will hold when $y > 1$. When $0 < y < 1$ we must take $k = -\frac{1}{2}$ and when $y < 0$ we must take $k = +\frac{1}{2}$.

Remembering that $p = \mathrm{d}y/\mathrm{d}x$, we find that this now further separates to give

$$k \,\mathrm{d}x = \frac{y-1}{y}\,\mathrm{d}y,$$

and so

$$kx = y - \log|y| + c. \tag{7.3}$$

We note that the solution (7.2) is included in equation (7.3) as the case $k = 0$.

Example 7.1.2

$$yy'' = y'(1 - y').$$

As before, we write $y' = p$ and $y'' = p\,(\mathrm{d}p/\mathrm{d}y)$, so that

$$yp\,\frac{\mathrm{d}p}{\mathrm{d}y} = p(1 - p). \tag{7.4}$$

Either $p = 0,$

that is, $y = \text{const.},$

or

$$y\,\frac{\mathrm{d}p}{\mathrm{d}y} = 1 - p\,;$$

therefore

$$\frac{\mathrm{d}p}{1 - p} = \frac{\mathrm{d}y}{y},$$

and

$$-\log|1 - p| = \log|y| + \text{const.}$$

Therefore

$$y(1 - p) = c,$$

and

$$p = \frac{\mathrm{d}y}{\mathrm{d}x} = \frac{y - c}{y}.$$

Finally

$$x = \int \frac{y}{y - c} \, dy$$

$$= \int \left\{ 1 + \frac{c}{y - c} \right\} dy$$

$$= y + c \log |y - c| + d. \qquad (7.5)$$

In this case the solution (7.4) is not included as a special case of (7.5).

Example 7.1.3

$$y'' + f(y)y' + g(y)(y')^2 = 0.$$

Here again, since x does not appear explicitly, we write

$$y' = p \quad \text{and} \quad y'' = p \frac{dp}{dy}.$$

Thus

$$p \frac{dp}{dy} + f(y)p + g(y)p^2 = 0.$$

The possibility

$$p = 0, \text{ that is, } y = \text{const.},$$

obviously arises. If this is not the case, division by p yields

$$\frac{dp}{dy} + g(y)p = -f(y),$$

which is a first-order linear equation for p in terms of y; this is readily solved as shown in Chapter 2.

Example 7.1.4

$$ay'' + by' + cy = 0.$$

It is worth noting that the procedure of this section also provides an alternative way of approaching the second-order linear equation with constant coefficients. Upon making the change of variable we have

$$ap \frac{dp}{dy} + bp + cy = 0;$$

this is a first-order homogeneous equation, which may also be solved by the methods of Chapter 2.

7.2 Equations of the type $F(y/x, y', xy'') = 0$

This type of equation is sometimes referred to as 'homogeneous', and the analogy with the first order homogeneous equations of Chapter 2 is fairly clear. It is reasonable to try the same transformation, namely

$$\frac{y}{x} = v, \qquad y' = p,$$

and this leads to the result

$$\frac{d}{dx} = \frac{d}{dv} \cdot \frac{dv}{dx} \tag{7.6}$$

$$= \frac{d}{dv} \cdot \left\{ \frac{1}{x} \frac{dy}{dx} - \frac{1}{x^2} y \right\},$$

so that

$$xy'' = \frac{dp}{dv}(p - v).$$

The original equation then becomes

$$F\left(v, p, \overline{p - v} \frac{dp}{dv}\right) = 0,$$

which is a first-order equation relating p and v. If this can be solved, the solution of the original equation follows from equation (7.6), which shows that

$$\int \frac{dx}{x} = \int \frac{dv}{p - v}. \tag{7.7}$$

Example 7.2.1

$$x^2(yy'' - y'^2) + y^2 = 0.$$

We note that, upon dividing by x^2, the equation may be written as

$$\frac{y}{x} \cdot xy'' - y'^2 + \left(\frac{y}{x}\right)^2 = 0,$$

so it is certainly of the type under consideration. After making the suggested change of variable we have

$$v(p - v)\frac{\mathrm{d}p}{\mathrm{d}v} = p^2 - v^2,$$

a first-order equation connecting p and v. To solve this, we note the two possibilities. Either

$$p - v = 0,$$

which leads to

$$y = Ax, \tag{7.8}$$

or, cancelling by $p - v$, we have

$$v\frac{\mathrm{d}p}{\mathrm{d}v} = p + v,$$

which is both linear and homogeneous. The solution is readily derived, and is

$$p = v \log|v| + cv.$$

It remains to use equation (7.7) to determine x and this yields

$$\begin{aligned}
\log|x| &= \int \frac{\mathrm{d}v}{p - v} \\
&= \int \frac{\mathrm{d}v}{v\{\log|v| + c - 1\}} \\
&= \log|\log|v| + c - 1| + \text{const.}
\end{aligned}$$

Upon taking exponentials, this becomes

$$|x| = k\{\log|v| + c - 1\}$$

so

$$x = B \log v + C \text{ say.} \tag{7.9}$$

Again taking exponentials, equation (7.9) may be written as

$$v = \alpha\, \mathrm{e}^{\beta x},$$

and hence

$$y = \alpha x\, \mathrm{e}^{\beta x}. \tag{7.10}$$

The solution (7.8) corresponds to the special case $\beta = 0$.

7.3   Equations of the type $F\{(y/x^r), (y'/x^{r-1}), (y''/x^{r-2})\} = 0$

This is clearly a generalization of the equation considered in the previous section which corresponds to the case $r = 1$. In this case it is natural to make a change of variable

$$y = x^r v.$$

Thus

$$y' = x^r v' + rx^{r-1} v,$$

and

$$y'' = x^r v'' + 2rx^{r-1} v' + r(r-1) x^{r-2} v.$$

Accordingly

$$\frac{y}{x^r} = v,$$

$$\frac{y'}{x^{r-1}} = xv' + rv,$$

$$\frac{y''}{x^{r-2}} = x^2 v'' + 2rxv' + r(r-1)v,$$

and our equation takes the form

$$F\{v, xv' + rv, x^2 v'' + 2rxv' + r(r-1)v\} = 0,$$

or

$$G(v, xv', x^2 v'') = 0. \tag{7.11}$$

This equation may be further simplified by a change of the independent variable. We write

$$x = e^t,$$

so that

$$x \frac{dv}{dx} = \frac{dv}{dt},$$

and

$$x^2 \frac{d^2 v}{dx^2} = \frac{d^2 v}{dt^2} - \frac{dv}{dt}.$$

The equation (7.11) therefore takes the form

$$G\left(v, \frac{dv}{dt}, \frac{d^2v}{dt^2} - \frac{dv}{dt}\right) = 0,$$

or

$$H\left(v, \frac{dv}{dt}, \frac{d^2v}{dt^2}\right) = 0.$$

This is now an equation in which the independent variable t is not explicitly present, and which can therefore be reduced to a first-order equation by means of the substitution discussed in Section 7.1.

Although this approach is more generally valid than that of Section 7.2, since it is valid for all r, it is clear that the method of Section 7.2 is to be preferred when $r = 1$, by virtue of its greater simplicity in requiring only one change of variable.

Example 7.3.1

$$xy'' + 2y' = (xy')^2 - y^2.$$

We begin by asking whether the equation takes the form

$$F\left(\frac{y}{x^r}, \frac{y'}{x^{r-1}}, \frac{y''}{x^{r-2}}\right) = 0 \qquad (7.12)$$

for any value of r. If we divide by x^{2r}, we have

$$\frac{y''}{x^{2r-1}} + 2\frac{y'}{x^{2r}} = \left(\frac{y'}{x^{r-1}}\right)^2 - \left(\frac{y}{x^r}\right)^2. \qquad (7.13)$$

This is of the form (7.12) provided that $2r = r - 1$ and $2r - 1 = r - 2$, which both hold if $r = -1$.

We accordingly make the changes of variable

$$y = \frac{1}{x}v, \; x = e^t, \qquad (7.14)$$

whence the equation (7.13) takes the form

$$\left(\frac{d^2v}{dt^2} - 3\frac{dv}{dt} + 2v\right) + 2\left(\frac{dv}{dt} - v\right) = \left(\frac{dv}{dt} - v\right)^2 - v^2,$$

which simplifies to

$$\frac{\mathrm{d}^2 v}{\mathrm{d}t^2} - \frac{\mathrm{d}v}{\mathrm{d}t} = \left(\frac{\mathrm{d}v}{\mathrm{d}t}\right)^2 - 2v\frac{\mathrm{d}v}{\mathrm{d}t}. \tag{7.15}$$

This, as anticipated, does not involve t explicitly, so we take $(\mathrm{d}v/\mathrm{d}t) = p$ as the new dependent variable, v as independent variable. Equation (7.15) then becomes

$$p\frac{\mathrm{d}p}{\mathrm{d}v} - p = p^2 - 2vp.$$

Either $p = 0$, that is, $v = \text{const.}$, that is, $y = c/x$, or, dividing by p,

$$\frac{\mathrm{d}p}{\mathrm{d}v} = 1 + p - 2v.$$

This may be solved as a linear equation for p in terms of v, and after the appropriate manipulation it is found that the solution is

$$p = A\,\mathrm{e}^v + 1 + 2v.$$

Finally, we have

$$\log x = t = \int \frac{\mathrm{d}v}{p} + \text{const.}$$

$$= \int_0^v \frac{\mathrm{d}v}{A\mathrm{e}^v + 1 + 2v} + B. \tag{7.16}$$

Equation (7.14) shows that

$$y = \frac{1}{x}v. \tag{7.17}$$

Equations (7.16) and (7.17) then relate y and x, with v as a parameter.

7.4 Equations of the type $F\{x, (y'/y), (y''/y)\} = 0$

It will be clear that the types of equation already considered in this chapter either do not explicitly involve the independent variable x (Section 7.1) or involve it only via powers of x (Sections 7.2 and 7.3). The type now under discussion, although more restrictive in the way y, y', and y'' arise, is less restrictive in that x can appear in any way whatsoever.

We introduce a new variable

$$v = \frac{y'}{y}, \tag{7.18}$$

so that

$$v' = \frac{y''}{y} - \left(\frac{y'}{y}\right)^2,$$

that is,

$$\frac{y''}{y} = v' + v^2. \tag{7.19}$$

Thus the equation

$$F\left(x, \frac{y'}{y}, \frac{y''}{y}\right) = 0$$

takes the form

$$F(x, v, v' + v^2) = 0,$$

which is a first-order equation for v in terms of x.

Example 7.4.1

$$yy'' - y'^2 + y^2 f(x) = 0.$$

It may be noted that Example 7.2.1 is the special case for which $f(x) = x^{-2}$. Since y, y', y'', appear only in second degree terms, we divide by y^2 to obtain

$$\frac{y''}{y} - \left(\frac{y'}{y}\right)^2 + f(x) = 0.$$

Upon making the substitution (7.18), and by virtue of equation (7.19), this becomes

$$(v' + v^2) - v^2 = -f(x),$$

that is,

$$v' = -f(x),$$

so

$$v = \frac{1}{y}\frac{dy}{dx} = A - \int f(x)\,dx.$$

Therefore

$$\log |y| = Ax - \int \{\int \int f(x)\, dx\}\, dx + B. \qquad (7.20)$$

For the special case in which $f(x) = x^{-2}$, it follows that

$$\int f(x)\, dx = -x^{-1},$$

and

$$\int \{\int \int f(x)\, dx\}\, dx = -\log |x|,$$

so equation (7.20) then becomes

$$\log |y| = Ax + B + \log |x|,$$

or

$$y = B_1 x\, e^{Ax},$$

which is precisely the result obtained previously.

It hardly needs to be stressed that this particular example worked out so very simply because y'' appeared in the equation as part of the specific combination of terms $yy'' - y'^2$, and in general things are likely to be more complicated.

Example 7.4.2

$$y'y'' - y^2 \sin x = 0.$$

As before we substitute from (7.18) and (7.19), whence

$$v(v' + v^2) - \sin x = 0,$$

or

$$v' = \frac{\sin x}{v} - v^2.$$

This is a fairly straightforward first-order equation between v and x; although it cannot be integrated in terms of the elementary functions, a numerical solution presents no difficulties once the boundary conditions are known.

7.5 Linear equations

The most general second-order linear equation may be written as

$$a(x)y'' + b(x)y' + c(x)y = Q(x),$$

or, after dividing through by $a(x)$,

$$y'' + p(x)y' + q(x)y = f(x). \qquad (7.21)$$

Basically, we can always obtain the general solution of this equation by adding together the complementary function and a particular integral. We shall therefore begin our study of equation (7.21) by considering the CF, which is the general solution of the equation

$$y'' + p(x)y' + q(x)y = 0. \qquad (7.22)$$

Now we have already seen, in the special case when the coefficients $p(x)$, $q(x)$, are constants, that the general solution can always be obtained provided that *any* solution of the equation is known. Suppose, then, in this more general case, that $y = y_1$ is *any* solution of equation (7.22). We write

$$y = y_1 . u,$$

so that

$$y' = y_1 u' + y_1' u,$$

and

$$y'' = y_1 u'' + 2y_1' u' + y_1'' u.$$

Substitution into equation (7.22) yields

$$y_1 u'' + (2y_1' + py_1)u' + (y_1'' + py_1' + qy_1)u = 0,$$

or

$$y_1 u'' + (2y_1' + py_1)u' = 0, \qquad (7.23)$$

the coefficient of u being zero since y_1 is a solution of equation (7.22). The equation (7.23) is easily integrated to obtain u'. We write

$$\frac{u''}{u'} = -\frac{2y_1'}{y_1} - p,$$

so

$$\log |u'| = -2 \log |y_1| - \int^x p(x)\, dx + \text{const.},$$

and hence

$$u' = \frac{A}{y_1^2} W(x), \qquad (7.24)$$

where

$$W(x) = \exp \left\{ - \int^x p(x)\, dx \right\}. \tag{7.25}$$

We note, in passing, that we may take any lower limit of integration we wish in equation (7.25), as the resulting change of $W(x)$ by a constant factor can be incorporated into the arbitrary constant A in equation (7.24).

Equation (7.24) now integrates to give

$$u = A \int^x \frac{W(x)}{y_1^2}\, dx + B,$$

and so

$$y = Ay_1 \int^x \frac{W(x)}{y_1^2}\, dx + By_1, \tag{7.26}$$

where again the lower limit of integration can be chosen as desired, the arbitrary multiple of y being readily incorporated into By_1. The solution (7.26) has two independent arbitrary constants, and so is the general solution of equation (7.22).

Turning now to the non-homogeneous equation (7.21), we again write

$$y = y_1 u,$$

where y_1 is any solution of the homogeneous equation (7.22). Substitution yields

$$y_1 u'' + (2y_1' + py_1)u' + (y_1'' + py_1' + qy_1)u = f,$$

that is,

$$u'' + \left(\frac{2y_1'}{y_1} + p \right) u' = \frac{f}{y_1}, \tag{7.27}$$

as before. This linear equation for u' has an integrating factor

$$\exp \left\{ \int \left(\frac{2y_1'}{y_1} + p \right) dx \right\} = \frac{y_1'^2}{W},$$

and, after multiplying through by this factor, equation (7.27) becomes

$$\frac{d}{dx} \left(\frac{y_1^2}{W} u' \right) = \frac{y_1 f}{W}.$$

Hence

$$\frac{y_1^2}{W}u' = A + \int^x \frac{y_1 f}{W}\,dx,$$

$$u' = A\frac{W}{y_1^2} + \frac{W}{y_1^2}\int^x \frac{y_1 f}{W}\,dx,$$

and

$$u = B + A\int^x \frac{W}{y_1^2}\,dx + \int^x \frac{W}{y_1^2}\left\{\int^x \frac{y_1 f}{W}\,dx\right\}dx.$$

Finally

$$y = By_1 + Ay_1\int^x \frac{W}{y_1^2}\,dx + y_1\int^x \frac{W}{y_1^2}\left\{\int^x \frac{y_1 f}{W}\,dx\right\}dx.$$

Again we stress that the lower limits of integration may be chosen arbitrarily, but are not additional independent arbitrary constants.

We have thus seen that we can always find the general solution of the equation (7.21) provided that any solution of the homogeneous equation (7.22) is available. It may well be asked how such a solution can be known in practice. If need be, of course, it can be obtained numerically—since *any* solution will do, we choose one with boundary conditions leading to an easily obtained solution. It also occasionally happens that there is a particularly simple solution of equation (7.22) which can be found by inspection.

Example 7.5.1

$$y'' + xy' - y = x^2.$$

Now it is fairly easily seen that the homogeneous equation

$$y'' + xy' - y = 0$$

has a solution

$$y = x.$$

We accordingly write

$$y = x.v$$

and substitute into the full equation. This yields

$$(xv'' + 2v') + x(xv' + v) - xv = x^2,$$

or

$$v'' + \left(x + \frac{2}{x}\right)v' = x. \tag{7.28}$$

The integrating factor is

$$\exp\left\{\int^x \left(x + \frac{2}{x}\right) dx\right\} = x^2 \exp\left(\tfrac{1}{2}x^2\right).$$

Thus equation (7.28) becomes

$$\frac{d}{dx}\{v'x^2 \exp\left(\tfrac{1}{2}x^2\right)\} = x^3 \exp\left(\tfrac{1}{2}x^2\right),$$

so that

$$v'x^2 \exp\left(\tfrac{1}{2}x^2\right) = A + \int^x x^3 \exp\left(\tfrac{1}{2}x^2\right) dx$$

$$= A + (x^2 - 2) \exp\left(\tfrac{1}{2}x^2\right).$$

Therefore

$$v' = Ax^{-2} \exp\left(-\tfrac{1}{2}x^2\right) + 1 - \frac{2}{x^2},$$

$$v = A \int \frac{\exp\left(-\tfrac{1}{2}x^2\right)}{x^2} dx + x + \frac{2}{x} + B,$$

and

$$y = Ax \int^x \frac{\exp\left(-\tfrac{1}{2}x^2\right)}{x^2} dx + x^2 + 2 + Bx. \tag{7.29}$$

It is perhaps worth noting, finally, that

$$\int^x \frac{\exp\left(-\tfrac{1}{2}x^2\right)}{x^2} dx = \text{const.} - \frac{\exp\left(-\tfrac{1}{2}x^2\right)}{x} - \int^x \exp\left(-\tfrac{1}{2}x^2\right) dx,$$

after integration by 'parts', so the result (7.29) can be alternatively expressed as

$$y = A_1\{\exp\left(-\tfrac{1}{2}x^2\right) + x \int^x \exp\left(-\tfrac{1}{2}x^2\right) dx\} + x^2 + 2 + B_1 x.$$

Example 7.5.2

$$x^2y'' + xy' + 2y = x^2.$$

It may be noted that this is an equation in which the CF can easily be obtained explicitly, by noting that the homogeneous equation

$$x^2y'' + 4xy' + 2y = 0 \qquad (7.30)$$

is of the type discussed in Section 7.3, with $r = 0$. Accordingly the change of independent variable

$$x = e^t$$

will reduce the equation to a form in which x (or t) does not appear explicitly, that is to a second-order linear equation with constant coefficients and this can be solved by the methods of Chapter 5.

Alternatively, the essential homogeneity of equation (7.30) leads us to look for a solution of the form

$$y \propto x^n,$$

which will hold provided that

$$n(n - 1) + 4n + 2 = 0,$$

that is,

$$n^2 + 3n + 2 = 0,$$

that is,

$$n = -1 \text{ or } -2.$$

We accordingly write

$$y = x^{-1}v,$$

whence the full equation becomes

$$x^2(x^{-1}v'' - 2x^{-2}v' + 2x^{-3}v) + 4x(x^{-1}v' - x^{-2}v) + 2x^{-1}v$$
$$= x^2,$$

that is,

$$xv'' + 2v' = x^2,$$

or

$$v'' + \frac{2}{x}v' = x. \qquad (7.31)$$

The integrating factor is

$$\exp\left(\int \frac{2}{x}\,dx\right) = x^2,$$

so equation (7.31) becomes

$$\frac{d}{dx}(v'x^2) = x^3.$$

Therefore

$$v'x^2 = A + \tfrac{1}{4}x^4,$$
$$v' = Ax^{-2} + \tfrac{1}{4}x^2,$$
$$v = -Ax^{-1} + \tfrac{1}{12}x^3 + B,$$

and

$$y = \tfrac{1}{12}x^2 + \frac{A_1}{x^2} + \frac{B}{x}.$$

Tutorial Examples

1. Obtain the general solutions of the following differential equations.
   (a) $yy'' = (y')^2(1 - y')$.
   (b) $y'' + \tanh y(y')^2 + \operatorname{sech}^2 y = 0$.

2. A particle of mass m is projected at time $t = 0$ from a position $x = 0$ with a speed $v = v_0$. It moves in a straight line along a horizontal plane under the action of a frictional resistance equal to

$$mk_1v + mk_2v^2,$$

where k_1 is a constant but k_2 is a known function of x. Set up the equation of motion for the particle, and show that

$$vF'(x) = v_0 - k_1F(x) = v_0\,e^{-k_1t},$$

where

$$F(x) = \int_0^x \exp\left\{\int_0^z k_2(\xi)\,d\xi\right\}dz.$$

3. Solve the equation

$$yy' + xy''(xy' + y) = 0,$$

(a) as an equation of the form $F\{(y/x), y', xy''\} = 0$,
(b) as an equation of the form $F(y, xy', x^2y'') = 0$.

4. Solve the equation

$$x^3y'(yy'' - y'^2) + y^3 = 0,$$

(a) as an equation of the form $F\{(y/x), y', xy''\} = 0$,
(b) as an equation of the form $F\{x, (y'/y), (y''/y)\} = 0$.

5. Solve the equation

$$x^2y'' + y = x^2,$$

(a) as an equation of the form $F(y, xy', x^2y'') = 0$,
(b) as a linear equation.

6. Solve the following linear equations.

(a) $(1 + x^2)y'' - 2y = x$.
(b) $y'' + n^2y = f(x)$.
(c) $x^2y'' - 4xy' + 6y = x^5 \sin x$.

Answers to Tutorial Examples

Chapter 1

1. (a) 1st order, linear.
   (b) 3rd order, linear.
   (c) 2nd order, linear.
   (d) 2nd order, non-linear, 1st degree.

2. (a) $y''(1 + \cot x) + 2y' + y(1 - \cot x) = 0$;
      2nd order, 1st degree.
   (b) $2xyy' - y^2 + x = 0$; 1st order, 1st degree.
   (c) $y''' - 2y'' + y' + 2y = 0$; 3rd order, 1st degree.

Chapter 2

1. (a) $y^2 = cx^2 - 1$.
   (b) $\exp(x + y) = c(x + 2)^2 (y + 2)^2$.
   (c) $r = c(1 - b \cos \theta)$.

2. (a) $x = \tan \dfrac{x + y}{2} + c$.

   (b) $x^2 + y^2 = c\,e^{2x}$.
   (c) $xy^2 = c - \cos x$.

3. (a) $x = \tan\left(t + \dfrac{\pi}{4}\right) = \dfrac{\tan t}{1 - \tan t}$.

   (b) $x = 1 + t + \frac{1}{3}t^3$.
   (c) $\log x = \frac{1}{2}t^2 - 2$.
   (d) $x = \log t + 2$.

4. (a) $x^4 + 2x^2y^2 = c$.
   (b) $x^2 + 2xy = c$.
   (c) $x^2 + xy - y^2 = c$.
   (d) $2y = c - x^2/c$.

5. (a) $(x - 5y + 5)(x - y - 1) = c$.
   (b) $x = c - \frac{8}{5} \log |5x + 10y + 17|$.

6. (a) $y = \operatorname{sech} x(c + \tanh x)$.
   (b) $y = \frac{1}{2} \sin x + (\frac{1}{2}x + c) \sec x$.
   (c) $y = \frac{1}{2}(1 - x) + (1 - x^2)\left\{c + \frac{1}{4} \log \left|\frac{1 + x}{1 - x}\right|\right\}$.

7. (a) $x^2 + 3xy + y^3 = c$.
   (b) $xy + y^2 = \sin x + c$.

8. (a) $3xy^2 - y^3 = c$.
   (b) $x^2 - y^2 - 2xy + 2x - 2y = c$.
   (c) $xy(x + 3) = c$.

9. (a) $y = x + \dfrac{1 + ce^{2x}}{1 - ce^{2x}}$.

   (b) $\sin y = cx \exp (\frac{1}{2}x^2)$.
   (c) $4 \log |1 - 3x + 6y| = 6x - 3y + c$.
   (d) $x^3y^3\{c - 3 \log |x|\} = 1$.
   (e) $y^2 = \cosh^2 x\{2x \tanh x - 2 \log \cosh x + c\}$.
   (f) $x = \cosh y\{c - \tanh y\}^{-1}$,

10. $\sin x + \tan x \tan y = c$.

Chapter 3

1. $x = y' + \log |y'| + c$.
2. $y = cx + (1 + c^2)^{\frac{1}{2}}$; $x^2 + y^2 = 1$.
3. $y = x + \frac{1}{3}$; $9(c - y)^2 = 4(c - x)^3$.
4. $y = 1 + x\{c - \log |x|\}^{-1}$.

Chapter 4

4. $2kx_m = \log (1 + kv_0^2/g)$.
   $v_1 = v_0(1 + kv_0^2/g)^{-\frac{1}{2}}$.

Chapter 5

1. (a) $y = A e^{-3x} + B e^{-4x}$.
   (b) $y = (A + Bx) e^{-2x}$.
   (c) $y = e^{-\frac{1}{8}x}\{A \cos \frac{1}{8}x\sqrt{7} + B \sin \frac{1}{8}x\sqrt{7}\}$.
   (d) $y = A e^x + B e^{-2x}$.

2. (a) $y = \frac{1}{5}x^2 - \frac{4}{25}x + \frac{23}{125}$.

   (b) $y = \frac{1}{18}x^3 + \frac{1}{18}x^2 - \frac{2}{27}x$.

   (c) $y = -\frac{1}{6}\cos 2x$.

   (d) $y = \frac{1}{4}x \sin 2x$.

   (e) $y = x\,e^{-x} - 2x\,e^{-2x}$.

3. (a) $y = \frac{1}{4}x^2 - \frac{3}{8}x + \frac{15}{32}$
   $+ \frac{1}{32}e^{-\frac{5}{4}x}\{17\cos\frac{1}{4}x\sqrt{7} + 19\sqrt{7}\sin\frac{1}{4}x\sqrt{7}\}$.

   (b) $y = \frac{1}{8}\cosh 2x + \dfrac{7 - \cosh\pi}{8\sinh\pi}\sinh 2x - \frac{1}{8}\cos 2x$
   $- \frac{1}{8}\sin 2x$.

   (c) $y = \frac{2}{9}e^x + \frac{1}{9}e^{-2x}\{16 - (8 + e^6)x\}$.

4. (a) $y = A\,e^x + B\,e^{-x} - 1 - e^{-x}\log(1 + e^x)$
   $+ e^x \log(1 + e^{-x})$.

   (b) $y = A\,e^{-\frac{5}{3}x} + B\,e^{-x} + \frac{1}{14}e^x(x^2 - \frac{13}{7}x + \frac{127}{98})$.

Chapter 6

1. $kt = \dfrac{\pi}{3} + \frac{1}{2}\log(2 + \sqrt{3})$.

5. $\left(\dfrac{A}{P_0}\right)^2 = \dfrac{k^2 + n^2 D^2}{(k - mn^2)^2 + n^2 D^2}$,
   < 1 if $n^2 > 2k/m$.

9. $I = \dfrac{E_0}{R}\left\{\cos\omega_0 t - \dfrac{4b\varepsilon}{(9 + 16b^2)^{\frac{1}{2}}}\sin(2\omega_0 t + \alpha)\right\}$,
   where $\tan\alpha = \frac{4}{3}b$, and $b = \frac{1}{2}R(C/L)^{\frac{1}{2}}$.

Chapter 7

1. (a) $x = y + A\log|y| + B$, or $y = c$.

   (b) $x = \displaystyle\int_0^y \dfrac{\cosh y}{(A - 2y)^{\frac{1}{2}}}\,dy + B$.

3.†(a) $x = \dfrac{A}{p}\exp\sqrt{(c - 2\log|p|)}$,
   $y = A\sqrt{(c - 2\log|p|)}\exp\sqrt{(c - 2\log|p|)}$.

† Although the answers (a) and (b) look terribly different, they are equivalent. Verifying this fact is an amusing little exercise!

(b) $y = p \log |p| + Ap,$

$\log x = (A + 1) \log |p| + \frac{1}{2}\{\log |p|\}^2 + B.$

4. $y = ax \dfrac{\exp \sqrt{(1 + x^2/b^2)}}{1 + \sqrt{(1 + x^2/b^2)}}.$

5. $y = x^{\frac{1}{2}}\{A \cos (\frac{1}{2}\sqrt{3} \log x) + B \sin (\frac{1}{2}\sqrt{3} \log x)\} + \frac{1}{3}x^2.$

6. (a) $y = \frac{1}{2}(1 + c)(1 + x^2) \tan^{-1}x + \frac{1}{2}cx.$

(b) $y = A \cos nx + B \sin nx + \dfrac{1}{n} \displaystyle\int_0^x \sin n(x - t) f(t) \, \mathrm{d}t.$

(c) $y = Ax^3 + Bx^2 - 2x^2 \cos x - x^3 \sin x.$